“互联网+”教材系列
中国地质大学(武汉)实验教学系列教材
中国地质大学(武汉)实验教材项目(SJC-201908)资助
中国地质大学(武汉)国际学生地质实践教材基金资助

GEOLOGICAL FIELDTRIP GUIDEBOOK AT THE THREE GORGES REGION

主编　喻建新　冯庆来　王　岸　林启祥
译者　赖小春　汪卫红　周诗羽　刘倩倩　许　峰

CHINA UNIVERSITY OF GEOSCIENCES PRESS

图书在版编目(CIP)数据

三峡地区地质学实习指导书＝GEOLOGICAL FIELDTRIP GUIDEBOOK AT THE THREE GORGES REGION：英文/喻建新等主编；赖小春等译．—武汉：中国地质大学出版社，2023.1

"互联网＋"教材系列　中国地质大学(武汉)实验教学系列教材

ISBN 978-7-5625-5553-7

Ⅰ.①三…　Ⅱ.①喻…　②赖…　Ⅲ.①三峡-区域地质学-高等学校-教材-英文　Ⅳ.①P562

中国国家版本馆 CIP 数据核字(2023)第 059346 号

GEOLOGICAL FIELDTRIP GUIDEBOOK AT THE THREE GORGES REGION

喻建新　冯庆来　王　岸　林启祥　**主编**

赖小春　汪卫红　周诗羽　刘倩倩　许　峰　**译者**

责任编辑：龙昭月　　　　责任校对：何澍语

出版发行：中国地质大学出版社(武汉市洪山区鲁磨路 388 号)　　　　邮编：430074

电　　话：(027)67883511　　传真：(027)67883580　　E-mail：cbb@cug.edu.cn

经　　销：全国新华书店　　　　http://cugp.cug.edu.cn

开本：787 毫米×1092 毫米　1/16　　　　字数：440 千字　印张：14.25

版次：2023 年 1 月第 1 版　　　　印次：2023 年 1 月第 1 次印刷

印刷：武汉市籍缘印刷厂

ISBN 978-7-5625-5553-7　　　　定价：42.00 元

本书编委会

主　编:喻建新　冯庆来　王 岸　林启祥

译　者:赖小春　汪卫红　周诗羽　刘倩倩　许　峰

编　委:(按姓氏笔画排序)

王永标　王国庆　王家生　边秋娟　刘　嵘

汤华云　杜远生　杨江海　何卫红　余文超

张克信　徐　珍　徐亚东　彭　念　彭松柏

曾佐勋　楚道亮　廖群安　等

序

随着“一带一路”工作的实施和推进，国际交流与国际合作的机会日益增多。在当今全球化的背景下，高等教育国际化已成为大趋势。高等教育国际化将促进我国课程多样化，带来更丰富的地学教学资源。这不仅使得国外的地球科学教材更容易引入我国，而且国内的本土教材也越来越多地朝着国际化方向发展。

Geological Fieldtrip Guidebook at the Three Gorges Region 在这一背景下应运而生，具有以下特点：

(1)探索了由成熟的本土教材到国际化教材的建设路径。本教材以中文版的《三峡地区地质学实习指导手册》(喻建新等，2016)为基础，在中国地质大学(武汉)地球科学学院和外国语学院共同的辛勤努力下实现了本土教材的国际化，顺应了国际化地质类技术人才的培养需求。

(2)建立了完善的三峡地区地质学实习体系。本教材将三峡地区的自然地理、研究历史、区域地质、野外教学路线、教学程序和教学要求等内容进行了有机结合，为三峡地区的地质实习和地质教学提供了完善的体系。该教材是以地球科学优势学科为核心，集基础性、专业性、系统性与前沿性于一体的野外实践教材。

(3)提供了立体化的教学与学习模式。三峡地区地质实习是一套成熟的地球科学教学模式，集教材预习、课堂教学和野外实习于一体。同时，为了最大程度地满足教师课堂教学和学生课下学习的需要，本教材还配备有相应的慕课——“三峡地质野外实践”，提供了课堂教学、野外实习和在线学习的立体化教学和学习模式。另外，本书还配有相应路线的讲解视频，以二维码的形式呈现在书中。

本教材共分为三大部分：

(1)INTRODUCTION。包括 GEOLOGICIAL OVERVIEW(绪论)和 REGIONAL GEOLOGY(区域地质)两章。绪论介绍了实习区的自然地理，研究历史，实习目的、任务和要求；区域地质包括区域地层与古生物、沉积岩与沉积作用、岩浆岩与岩浆作用、变质岩与变质作用、地质构造的内容。

(2)FIELDTRIP ROUTE。共 10 章，涉及 10 条地质实习路线。每条路线都包含以下内容：教学路线、教学任务及要点、线路内容及观察点、教学进程及注意事项、专题研究及思考问题。

这 10 条实习路线分别是：

路线一　新元古代南华纪地层观察

路线二　震旦纪—寒武纪地层和古生物观察

路线三　宜昌黄花场奥陶纪大坪期地层观察

路线四　宜昌王家湾上奥陶统赫南特阶全球界线层型剖面和点位观察

路线五　奥陶纪晚期—二叠纪地层和古生物观察

路线六　晚古生代二叠纪地层和古生物观察

路线七　中三叠世—中侏罗世地层序列观察

路线八　长阳清江构造地质和寒武纪—奥陶纪地层观察

路线九　仙女山断裂及相关构造

路线十　第四纪路线

(3)Assement。共1章,包括以下内容:实习目的及实习阶段划分、各阶段主要教学内容及教学要求、实习成绩评定。

本书是为地质、资源、环境、工程等相关专业学生服务的野外实习指导教材,旨在帮助学生在完成相关理论课程后进行实地实习。建议教学周期为一个野外实习周期,24～32个学时。该教材特别适用于来华留学的相关专业学生。同时,本教材也适用于从事地质相关工作的专业人士进行英语自学。此外,该教材可以与慕课“三峡地质野外实践”结合使用,实现课堂、野外实习和线上学习的有效结合。

本教材的编写团队由地球科学领域成果丰硕的资深教授和活跃在前沿科学研究领域的杰出青年教师组成,包括王岸、王永标、王国庆、王家生、冯庆来、边秋娟、刘嵘、杜远生、杨江海、何卫红、张克信、林启祥、徐亚东、彭松柏、喻建新、曾佐勋、廖群安(按姓氏笔画排序)。其翻译团队由外国语学院的中青年杰出教师构成,每位老师都有海外留学或工作经历,并致力于将英语教学研究与地球科学相结合,包括赖小春、汪卫红、周诗羽、刘倩倩和许峰。本教材的校对还得到了地质学专业老师及博士研究生的倾力相助,包括余文超、汤华云、王岸、楚道亮、纵端文、彭念、徐珍、范牧、林雯洁、李冰冰等人。

本教材的顺利出版离不开中国地质大学(武汉)地球科学学院、实验室与设备管理处、国际教育学院和外国语学院的大力支持,同时也感谢本教材顾问团队和编委会为本教材的编写提供的宝贵建议。

由于编写时间仓促和编者水平有限,本教材难免存在不尽如人意之处。我们真诚希望使用本教材的广大教师、学生和其他专业人士给予指正,以便我们能够及时更正和改进。

译　者

2023年1月

CONTENTS

Ⅰ **INTRODUCTION** ······ (1)
1 GEOLOGICAL OVERVIEW ······ (2)
1.1 Physical Geography ······ (2)
1.2 Research History ······ (4)
1.3 Internship Objectives, Tasks, and Requirements ······ (7)
2 REGIONAL GEOLOGY ······ (10)
2.1 Regional Strata and Palaeontology ······ (10)
2.2 Sedimentary Rocks and Sedimentation ······ (49)
2.3 Magmatic Rock and Magmatism ······ (60)
2.4 Metamorphic Rocks and Metamorphism ······ (82)
2.5 Geological Structure ······ (99)
Ⅱ **FIELDTRIP ROUTES** ······ (117)
3 ROUTE ONE: OBSERVING THE NEOPROTEROZOIC NANHUAIAN PERIOD STRATIGRAPHIC SEQUENCE ······ (118)
3.1 Teaching Route ······ (118)
3.2 Teaching Tasks and Requirements ······ (118)
3.3 Route Information and Observing Points ······ (119)
3.4 Teaching Process and Precautions ······ (125)
3.5 Focused Study and Reflections ······ (126)
4 ROUTE TWO: OBSERVING THE SINIAN-CAMBRIAN STRATIGRAPHY AND PALEONTOLOGY ······ (127)
4.1 Teaching Route ······ (127)
4.2 Teaching Tasks and Requirements ······ (127)
4.3 Route Information and Observing Points ······ (127)
4.4 Teaching Process and Precautions ······ (132)
4.5 Focused Study and Reflections ······ (133)

5 ROUTE THREE: OBSERVING THE ORDOVICIAN DAPINGIAN STRATA AT HUANGHUACHANG, YICHANG ······ (134)
5.1 Teaching Route ······ (134)
5.2 Teaching Tasks and Requirements ······ (134)
5.3 Route Information and Observing Points ······ (134)
5.4 Teaching Process and Precautions ······ (140)
5.5 Focused Study and Reflections ······ (141)
6 ROUTE FOUR:OBSERVING THE UPPER ORDOVICIAN HIRNANTIAN GSSP IN WANGJIAWAN, YICHANG ······ (142)
6.1 Teaching Route ······ (142)
6.2 Teaching Tasks and Requirements ······ (142)
6.3 Route Information and Observing Points ······ (142)
6.4 Section Description and Fossil Collection ······ (144)
6.5 Teaching Process and Precautions ······ (149)
6.6 Focused Study and Reflections ······ (150)
6.7 More Information about the Hirnantian GSSP ······ (150)
7 ROUTE FIVE: OBSERVING LATE ORDOVICIAN-PERMIAN STRATIGRAPHY AND PALEONTOLOGY ······ (156)
7.1 Teaching Route ······ (156)
7.2 Teaching Objectives and Requirements ······ (156)
7.3 Route Information and Observing Points ······ (156)
7.4 Teaching Process and Precautions ······ (161)
7.5 Focused Study and Reflections ······ (162)
8 ROUTE SIX: OBSERVING PERMIAN STRATIGRAPHY AND PALEONTOLOGY IN LATE PALEOZOIC ······ (163)
8.1 Teaching Route ······ (163)
8.2 Teaching Tasks and Requirements ······ (163)
8.3 Route Information and Observing Points ······ (163)
8.4 Teaching Process and Precautions ······ (168)
8.5 Focused Study and Reflections ······ (168)
9 ROUTE SEVEN: OBSERVING MIDDLE TRIASSIC-MIDDLE JURASSIC STRATIGRAPHY ······ (169)
9.1 Teaching Route ······ (169)
9.2 Teaching Tasks and Requirements ······ (169)
9.3 Route Information and Observing Points ······ (169)
9.4 Teaching Process and Precautions ······ (173)
9.5 Focused Study and Reflections ······ (174)

10 ROUTE EIGHT: OBSERVING STRUCTURAL GEOLOGY AND CAMBRIAN-ORDOVICIAN STRATA IN QINGJIANG, CHANGYANG …… (175)
10.1 Teaching Route …… (175)
10.2 Teaching Tasks and Requirements …… (175)
10.3 Route Information and Observing Points …… (175)
10.4 Teaching Process and Precautions …… (188)
10.5 Focused Study and Reflections …… (189)
11 ROUTE NINE: OBSERVING THE XIANNVSHAN FAULT AND RELATED STRUCTURES …… (190)
11.1 Teaching Route …… (190)
11.2 Teaching Tasks and Requirements …… (190)
11.3 Route Information and Observing Points …… (190)
11.4 Teaching Process and Precautions …… (194)
11.5 Focused Study and Reflections …… (195)
12 ROUTE TEN: THE QUATERNARY ROUTE …… (196)
12.1 Teaching Route …… (196)
12.2 Teaching Tasks and Requirements …… (196)
12.3 Route Information and Observing Points …… (196)
12.4 Teaching Process and Precautions …… (205)
12.5 Focused Study and Reflections …… (205)
Ⅲ ASSESSMENT …… (207)
13 ASSESSMENT OF TEACHING PROCEDURES AND PRACTICE PERFORMANCES …… (208)
13.1 Objectives and Stage Division …… (208)
13.2 Main Teaching Contents and Teaching Requirements …… (209)
13.3 Performance Assessment …… (212)
REFERENCES …… (213)

I

INTRODUCTION

1 GEOLOGICAL OVERVIEW

1.1 Physical Geography

The Three Gorges Zigui Research Base of China University of Geosciences (Wuhan) (Zigui Base for short) is located in the northwestern edge of Zigui County, about two kilometers away from the Three Gorges Dam. The base construction project was approved in 2002, and the construction began in 2004, with first phase of infrastructure construction having been completed in 2005. Various field practices and teaching activities have been carried out here since the year of 2006.

Zigui County is in the western part of Hubei Province, adjacent to Yichang City in the east, and about 400 kilometers away from Wuhan—the capital city of Hubei Province. The transport between Wuhan and Zigui is very convenient. Teachers and students can arrive in Yichang via the Hanyi Expressway or the high-speed railway from Wuhan to Yichang (dozens of shifts per day), and then go to Zigui via the highway between Yichang and Zigui. The shuttle bus between Yichang and Zigui travels every 15 minutes.

There are seven towns and six townships under the jurisdiction of Zigui County, namely Maoping, Quyuan, Guizhou, Shazhenxi, Lianghekou, Guojiaba, Yanglinqiao, and Shuitianba, Xietan, Moping, Meijiahe, Zhouping, Zhilan (Figure 1-1). The county currently has 202 administrative villages, seven resident committees, 1, 182 villager groups, and 43 resident groups. Zigui County covers a land area of 2, 427 km^2 with a population about 423, 000 in aggregate. In 2011, Zigui's regional GDP reached 6.7 billion yuan, an increase of 26.4% compared with the 5.3 billion yuan in 2010.

Zigui is rich in mineral resources, and more than 20 kinds of mineral ores have already been discovered in the county, including iron mine, gold mine, coal mine, limestone, barite, etc. In addition, the hydropower resources are adequate. Traversed by the Yangtze River, Zigui County presents huge hydropower potential. Zigui County, dotted with small- and medium-sized hydropower stations, has become China's primary electrification construction county and serves as the pilot county for intermediate hydropower construction in rural areas. The thermal power installed capacity is about 30, 000 kilowatts and the

Figure 1-1 The location of the Zigui area

annual power generation capacity can reach 180 million kW • h.

Zigui County has 23,900 hectares of land area, mostly deserted mountains and under-managed forests. It is a typical agricultural county in mountainous areas. In recent years, by vigorously developing diversified economies and market-oriented agriculture, the county has basically formed agricultural production bases of alpine flue-cured tobacco and off-season vegetables, Zhongshan tea and chestnut, and low mountain citrus, with high-output economic forest area of 280,000 mu. Rich in local agricultural products, Zigui is famous for citrus, tea, flue-cured tobacco, chestnut, and konjac. Navel orange, jin orange, taoye orange (peach leaf orange) and summer orange are known as "Four Specialties in Xiajiang", in particular, the navel orange is famous in China. The county's planting area for navel orange stands at 150,000 mu. Zigui County, with a large-scale navel orange planting area and high-quality navel orange, enjoys the fame of "the Hometown of Chinese navel orange" given by the Ministry of Agriculture and Rural Affairs of the People's Republic of China. Oranges here have won the gold medal for quality fruit and the name of "Chinese Famous Fruit" for many times.

The internship area is at the eastern end of Daba Mountains which are situated at the second gradient terrain of the Three Gradient Terrains of China. It falls into the mountainous area in the southwestern Hubei Province, and it is at the river valley in the lower section of the upper reaches of the Yangtze River. The direction of the mountain range is northeast-southwest or northwest-southeast. This area is in the subtropical monsoon zone with humid climate. With the high mountains and the cushion of water, the inversion layer is formed below 600 meters. That is to say that the warm winter zone is formed along the banks of the river in winter. The annual temperature averages 18 ℃ with the minimum

temperature at only about −3 ℃. The annual frost-free period is 306 days. The relative humidity of the air is 72% with annual rainfall at 1,016 mm. There is often heavy rain in summer, making the land vulnerable to flooding disasters and soil erosion.

1.2 Research History

The Huangling Dome area of the Three Gorges of the Yangtze River is one of the regions with a relatively long history and a high degree of research in China's regional geological survey. During 1863–1914, Raphael W. Pumpelly from the United States and Ferdinand von Richthofen from Germany made a rough geological survey in the Three Gorges area. In the 1920s, Li Siguang and Zhao Yazeng (1924), the major founders of modern geology in China, completed the stratigraphic geological survey of the Zigui–Yichang Section of the Three Gorges of the Yangtze River, which laid foundation for the strata-tectonic framework in this area. Later, famous geologists Xie Jiarong, Zhao Yazeng, Xu Jie, Yin Zanxun, Lu Yanhao, Zhang Wentang, etc., successively conducted more in-depth research, which paved the way for regional geology research.

More than ten institutes and departments have conducted overall geological surveys or mineral exploration work in this area since 1949. From the late 1950s to the early 1960s, Mr. Yang Zunyi led a group of teachers and students from Beijing College of Geology to conduct regional geological survey of West Yichang on the scale of 1/200,000. A systematic study was conducted on strata in all ages of the Three Gorges area. Since then, Regional Geological Survey Team of Hubei Province has carried out regional geological survey of East Yichang on the scale of 1/200,000, and in 1970 it was combined with East Yichang and Changyang for publication.

In the 1970s, Hubei Three Gorges Strata Team (1978, jointly built by Hubei Bureau of Geology and Mineral Resources, Hubei Geological Museum as well as Yichang Institute of Geology and Mineral Resources) and Nanjing Institute of Geology and Palaeonotology, Chinese Academy of Sciences (1978) had carried out in-depth studies on the Sinian–Permian strata in this area. In the 1980s, Yichang Institute of Geology and Mineral Resources, Institute of Geology and Minerals of the Ministry of Mines, and Hubei Institute of Geology, through the systematic and indepth study, successively published a series of studies on Sinian (赵自强等, 1985), Early Paleozoic (汪啸风等, 1987), Late Paleozoic (冯少南等, 1985), Triassic–Jurassic (张振来等, 1985), and Cretaceous–Tertiary (雷奕振等, 1987),

systematically studied and summarized the Sinian–Teritary stratigraphy and palaeontology in the Three Gorges area of the Yangtze River. Studies on lithostratigraphy, biostratigraphy, and chronostratigraphy reached the leading level domestically. The research results of the Sinian, the Sinian/Cambrian boundary, and the Ordovician/Silurian boundary were internationally advanced at that time. In order to cooperate with the urban development plan of Yichang City, Exi Geological Team of Hubei Province led the project from 1986 to 1990 to complete the 1/50,000 scale geological mapping of Yichang with the existing data. Subsequently, the 1/50,000 scale regional geological mapping of West Liantuo and West Sandouping were completed in 1991.

Since the mid-to-late 1990s, with the support of Ministry of Land and Resources of the People's Republic of China and the Bureau of Three Gorges Immigration under State Council, *The Protection of Precious Geological Heritage and the Archean–Mesozoic Stratigraphic Subdivision and Eustatic Sea-Level Change in Three Gorges* have been completed by Yichang Institute of Geology and Mineral Resources (汪啸风等, 2002). The results made up for the weak links in the stratigraphic sequence and the Archean–Mesoproterozoic study in the region, and further improved the study on the stratigraphic palaeontology, especially stratigraphic sequence and chronostratigraphy. During that time, Exi Geological Team completed a regional geological mapping of Fenxiangchang and East Liantuo on the scale of 1/50,000.

At the beginning of the 21st century, Ministry of Land and Resources rolled out a new round of land and resources surveys. Wuhan Geological Survey Center (formerly called Yichang Institute of Geology and Mineral Resources) of China Geological Survey, Institute of Geology, Chinese Academy of Geological Sciences (CAS), and Nanjing Institute of Geology and Paleontology, CAS have successively carried out a series of studies on biodiversity events in Sinian and the division of chronostratigraphy units, the division and correlation of chronostratigraphy units in Sinian and the Lower Palaeozoic in South China. Their studies on chronostratigraphy unit division and correlation in Sinian have further improved the internal chronostratigraphic system in Sinian (陈孝红等, 2002). Wuhan Geological Survey Center and Nanjing Institute of Geology and Paleontology respectively completed research on Global Stratotype Section and Points (GSSP, "Golden Spike") of Heirnantian (the Upper Ordovician in Wangjiawan, Yichang), and the Middle/Lower Ordovician and Dapingian (the third stage of Ordovician in Huanghuachang, Yichang), which have greatly promoted the study of the Ordovician chronostratigraphy in the region and in the world. In addition, institutions such as China University of Geosciences (Beijing) and Institute of Geology, CAS have also achieved significant results in the study of chronostratigraphy of Sinian in the Three Gorges area and published papers in international publications such as *Nature* and *Episodes*. It has aroused the attention of their international counterparts and greatly enhanced the role of Sinian section of this area in the worldwide re-division of the Ediacaran.

The Huangling Dome area of the Three Gorges of the Yangtze River is not only a highlight in the study of stratigraphy in China, but also a key area for the investigation and prevention of geological disasters in China. Many institutes such as Changjiang Water Resources Commission of the Ministry of Water Resources, Three Gorges Geological Team and Hubei Earthquake Administration, Hydrogeologic Team of Nanjiang, Sichuan, Hubei Bureau of Geology and Mineral Resources, and Wuhan Geological Survey Center conducted surveys and detailed investigations of regional hydrology, engineering geology, and disaster geology on the scales of 1/100,000, 1/200,000, and 1/500,000. Research reports were prepared at the same time. Significant progress has been made in geological surveys of mountain stability, rock collapses, and landslides. Additionally, the Wuhan Earthquake Team, the Second Hydrographic Team of Hubei Province, the Changban Seismic Station, and Hubei Earthquake Administration have systematically observed the activities of ruptures in Xiannvshan, Zhouping, and Tianyangping which are at the vicinity of the internship base since the 1970s. Changjiang Water Resources Commission and China University of Geosciences (Wuhan) have also conducted detailed research on the ruptures in this area. These surveys and research work greatly enriched field practice teaching work.

Since the 1990s, a large number of researchers, teachers, and students at universities and research institutes at home and abroad have carried out many great research projects on pre-Nanhuaian metamorphic basement, Neoproterozoic granite complex, and sedimentary strata since Nanhuaian in the Huangling Dome area, the Yangtze Craton, especially of the formation age and geological significance of Archean gray gneiss (TTG) in the northern part of the Huangling Dome (高山等, 1990; 马大铨, 1992); the age and geological tectonic significance of Paleoproterozoic tectonic-magmatic-metamorphic thermal events (凌文黎, 2000; Qiu et al., 2000; Zhang et al., 2006; 张少兵等, 2007; 郑永飞等, 2007; 熊庆等, 2008; 彭敏等, 2009; Yin et al., 2013); the genetic type and age of Huangling granite complex in Neoproterozoic (马大铨, 2002; 李志昌等, 2002; 李益龙等, 2007; Zhang et al., 2008, 2009; Wei et al., 2013; Zhao et al., 2013); the relationship between the discovery of cold spring carbonates in "cap dolostone" at the bottom of the Doushantuo Formation in Sinian and the snowball Earth event in Neoproterozoic (Jiang et al., 2003; 王家生等, 2005, 2012; Wang et al., 2008); studies on the Sinian and Cambrian oceans (McFadden et al., 2008; 朱茂炎, 2010; Ling et al., 2013); the discovery and identification of Neoproterozoic ophiolite in Miaowan and its geotectonic significance (彭松柏等, 2010; Peng et al., 2012); the age and genetic mechanism of the uplift of the Huangling Dome in the Mesozoic–Cenozoic (沈传波等, 2009; 刘海军等, 2009; Ji et al., 2013). Those significant new findings made the Huangling Dome area an highlight in the research of major scientific problems in

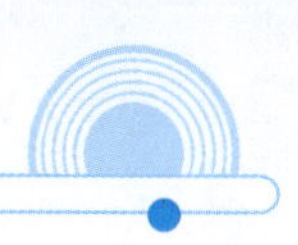

the frontier field of earth science, such as the growth and evolution of continental crust in the early Precambrian, the polymerization and break-up of Precambrian supercontinents (the Columbia supercontinent and the Rodina supercontinent), the origin and evolution of early life on Earth, the snowball Earth event in Neoproterozoic, the stretching and rifting of Mesozoic–Cenozoic. These new achievements provide an important scientific research foundation for re-recognizing the unique academic research status of the Huangling Dome in South China and even the world's geological tectonic evolution, and offer an important scientific research basis for the preparation of this internship manual.

This internship manual summarizes the overall research results of the predecessors and makes fine-tuned editing to suit undergraduate learning status quo. The assignments of each chapter and segment are as follows: Chapter 1, written by Feng Qinglai, translated by Zhou Shiyu, and proofread by Zhou Shiyu; Section 2.1, written by Lin Qixiang and Yu Jianxin, translated by Wang Weihong and Xu Feng, and proofread by Yu Wenchao, Section 2.2, written by Wang Guoqing, translated by Lai Xiaochun, and proofread by Yu Wenchao, Sections 2.3–2.5, written by Peng Songbai, Liao Qun'an, Zhou Hanwen, and Liu Rong, translated by Liu Qianqian, Xu Feng and Zhou Shiyu, and proofread by Tang Huayun and Wang An; Chapter 3, written by Peng Songbai, translated by Zhou Shiyu, and proofread by Chu Daoliang; Chapter 4, written by Peng Songbai and Liao Qun'an, translated by Zhou Shiyu, and proofread by Chu Daoliang; Chapter 5, written by Wang Jiasheng, translated by Wang Weihong, and proofread by Yu Wenchao; Chapter 6, written by Feng Qinglai, translated by Wang Weihong, and proofread by Chu Daoliang; Chapter 7, written by Zhang Kexin and Xu Yadong, translated by Liu Qianqian, and proofread by Yu Wenchao; Chapter 8, written by He Weihong, translated by Liu Qianqian, and proofread by Yu Wenchao; Chapters 9 and 10, written by Wang Yongbiao, translated by Zhou Shiyu and Lai Xiaochun, and proofread by Yu Wenchao and Wang An; Chapter 11, written by Yang Jianghai and Du Yuansheng, translated by Lai Xiaochun, and proofread by Wang An; Chapter 12, written by Zeng Zuoxun and Wang An, translated by Zhou Shiyu, and proofread by Chu Daoliang; Chapter 13, written by Wang Guoqing and Yu Jianxin, translated by Zhou Shiyu, and proofread by Zhou Shiyu.

1.3 Internship Objectives, Tasks, and Requirements

Field practice teaching is an essential and integral part for students majoring in geology.

The internship bases are set up to, on the one hand, stabilize the practical teaching team and provide logistics services for teachers and students; on the other hand, provide sustaining support to deepen the coordination between teaching and scientific research.

Since its establishment in 1952, China University of Geosciences (Wuhan) has paid great attention to field practice teaching and hands-on ability. Zhoukoudian and Beidaihe have always been the key field teaching practice bases for undergraduates majoring in geology in our school. These two areas are rich in various and abundant classical geological phenomena. However, for the current practical teaching, there are still some problems. First, with the development of natural resources, some classical and non-renewable geological phenomena have been damaged, thus severely affecting field teaching. Second, the two internship bases are located in North China. Our graduates have long lacked observation training about the geological effects and their geological records in South China, which has restrained students' understanding and knowledge of the geological phenomena in South China, thus affecting their future career. Third, the two bases have inadequate geological teaching resources in the sedimentary environment analysis and field identification of lithologic types of igneous rocks. To this end, undergraduates from China University of Geosciences (Wuhan) majoring in geology are required to take part in a two-week's field internship practice in Zigui to improve their fieldwork capability.

Based on the analysis of field teaching practice, the objectives of the field teaching in Zigui are as follows:

(1) By adding the field teaching practice in South China, students can observe the stratigraphic, lithologic, and tectonic characteristics there, and teachers will guide students to compare the similarities and differences between the stratigraphic sequences and the evolution laws of South China and North China.

(2) Students will have a well-placed grasp of the geological history development process in South China, so as to strengthen the weak links in the geological internship in North China, and thus develop their full-fledged geological knowledge and skills.

(3) Focused routes will be set. Students will carry out independent field surveys and collect geological data, so as to develop their ability to independently observe and analyze problems related to geological phenomena. That is how they can work to learn extensively with geology-science thinking patterns and geological working methods.

Zigui Base is rich in geological phenomena. To fulfill the teaching objectives mentioned above, ten teaching routes are selected and developed to carry out field teaching practice. Detailed information about each route is as follows.

Route One: Observing the Neoproterozoic Nanhuaian Period Stratigraphic Sequence

Route Two: Observing the Sinian–Cambrian Stratigraphy and Paleontology

Route Three: Observing the Ordovician Dapingian Strata at Huanghuachang, Yichang

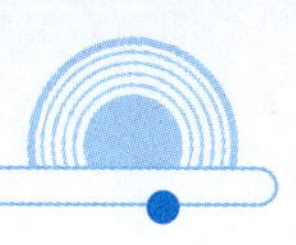

Route Four: Observing the Upper Ordovician Hirnantian GSSP in Wangjiawan, Yichang

Route Five: Observing Late Ordovician–Permian Stratigraphy and Paleontology

Route Six: Observing Permian Stratigraphy and Paleontology in Late Paleozoic

Route Seven: Observing Middle Triassic–Middle Jurassic Stratigraphy

Route Eight: Observing Structural Geology and Cambrian–Ordovician Strata in Qingjiang, Changyang

Route Nine: Observing the Xiannvshan Fault and Related Structures

Route Ten: The Quaternary Route

In order to ensure the quality of practical teaching and the orderly progress of the practice work, the teaching ideas of strict requirements and intensive training should always run through the whole practice process. The teaching approaches and methods can be adjusted in a flexible manner by the instructors in line with the basic teaching requirements. Basic requirements for field teaching on geological routes are as follows:

(1) On the day before field teaching, the instructor should inform the students of their teaching tasks, routes, objectives, requirements, and relevant precautions, so that they can prepare their thoughts, business, and equipment and items to carry.

(2) Instructor should check the number of students and their preparations before departure on a daily basis and at the end of each day's field teaching work. Instructor should carry out on-field check of the number of students, students' field records, specimens, samples as well as various instruments and equipment. Moreover, instructor should also give assignments and requirement to categorize their collected data. To deepen their understanding, students will be given some relevant questions to think and discuss in line with the content and requirements of the teaching route.

(3) In terms of teaching, instructor should first propose the teaching requirements, instead of talking alone, and then encourage the students to observe, discuss, and think, and finally record the procedure. This is how we train students to observe, describe, record, and collect information of geological phenomena in the field.

2 REGIONAL GEOLOGY

2.1 Regional Strata and Palaeontology

2.1.1 Introduction to the regional strata

The fieldwork site is located at Yichang, Hubei Province. Yichang belongs to the Upper Yangtze stratigraphic division of the South China Stratigraphic Superregion (湖北省地矿局, 1996). The strata in the area are well developed and typical for stratigraphic research at the Yangtze region, including standard stratotype sections of the Nanhuaian System of the Neoproterozoic and the Silurian system of the Lower Paleozoic and two "Golden Spikes". Here, strata of the Proterozoic, Palaeozoic and Mesozoic–Cenozoic were outcropped in turn. Among them, studies on the strata between the Neoproterozoic and the Lower Palaeozoic are best known. Strata since the Late Triassic show continental facies. Table 2-1 shows the main stratigraphic units in this region.

2.1.2 Introduction to strata in the fieldwork area

The fieldwork site is located at Zigui, Yichang, Hubei Province. At the site, most strata of the Yichang area could be seen. The following are introductions to some major ones (Table 2-1).

2.1.2.1 *The Proterozoic*

The strata outcropped at the fieldwork site are mainly of the Neoproterozoic. There are also many ourcrops of the Mesoprotozoic, but few of the Paleoproterozoic, except a few gneisses, schists, plagioclase amphibolites, quartzites, and marbles of the Shuiyuesi Group in Yichang and Xingshan County. In the guidebook, the Paleoproterozoic will not be discussed.

1. The Mesoprotozoic Kongling Rock Group (Pt_2*K.*)

The name of the Mesoprotozoic Kongling Rock Group (Pt_2*K.*) is evolved from

Table 2-1 Straigraphic sequence in the Three Gorges area

Chronostratigraphic Unit				Rock–Stratigraphic Unit			Code	Thickness/m	Lithological Description
Erathem	System	Series	Stage	Group	Formation	Member			
Cenozoic	Quaternary	Holocene		Qh				0–50	gravel, sand gravel, and sandy clay
		Pleistocene		Qp_3				>15	gravel layer, black clayey sand and yellowish-brown sandy clayey soil
				Qp_2				102	gravel layer, purplish-red gravel-bearing sandy clay and maroon reticulated clay
				Qp_1				21–27	gravel layer, yellowish-brown, and brownish-yellow siltstone with clayey siltstone
	Paleogene	Eocene			Pailoukou		E_2p	323–962	grayish-yellow to light purplish-red thick-bedded sandstone in the bottom, mainly composed of sandstone with fine-grained sandstone and mudstone
					Yangxi		E_2y	100–520	a set of grayish-brown, light red, and grayish-white medium–thick limestone under grayish-white, purplish-red thin–medium-bedded sandy limestone with variegated mudstone
		Paleocene			Gongjiachong		E_1g	60–470	brownish-red thick–massive breccia, conglomerate, or glutenite in the bottom, purplish-red mudstone and siltstone with brownish-yellow, brownish-red, and grayish-white sandstone, and grayish-green mudstone in the middle and top
Mesozoic	Cretaceous	Upper			Paomagang		K_2p	170–890	variegated sandstone, siltstone, silty mudstone, and mudstone in brownish-yellow, grayish-green, and yellowish-green
					Honghuatao		K_2h	773	bright reddish-brown thick-bedded sandstone with argillaceous fine-grained sandstone, siltstone, and mudstone
					Luojingtan		K_2l	400–600	purplish-red and gray thick-bedded–massive conglomerate, with the upper part intercalated with sandy conglomerate and gravel-bearing sandstone

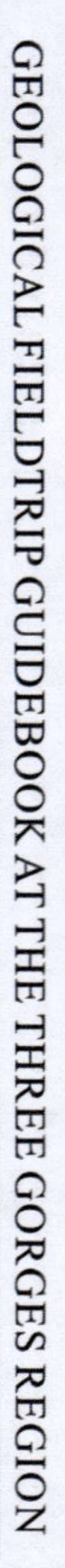

Table 2–1

Chronostratigraphic Unit				Rock–Stratigraphic Unit			Code	Thickness/m	Lithological Description
Erathem	System	Series	Stage	Group	Formation	Member			
Mesozoic	Cretaceous	Lower			Wulong		K_1w	714–1867	purplish-red and brownish-red medium–thick-bedded sandstone and gravel-bearing sandstone with conglomerate and argillaceous sandstone
					Shimen		K_1s	185–275	purplish-red and purplish-gray massive medium–coarse-grained conglomerate with red fine-grained sandstone lens
	Jurassic	Upper			Penglaizhen		J_3p	2115	purplish-grey feldspar quartz sandstone interbedded with argillaceous in different thicknesses, with yellowish-green shale and bioclastic limestone, containing fossils of ostracods, phyllopods, charophytes, and bivalves
					Suining		J_3s	630	purplish-red mud (shale) rocks with lithic arkose and siltstone, containing fossils of ostracods, contouropodas, and bivalves
		Middle			Shaximiao		J_2sh	1986	yellowish-grey and purplish-grey feldspar quartz sandstone interbedded with purplish-red and purplish-grey argillaceous (shale) rocks in different thicknesses
					Qianfoya		J_2q	390	purplish-red and greenish-yellow mudstone, siltstone, and fine-grained quartz sandstone with shell limestone
		Lower		Xiangxi	Tongzhuyuan		J_1t	280	yellow, yellowish-green, and grayish-yellow sandy shale, siltstone, and feldspar quartz sandstone with carbonaceous shale and thin coal layer or coal line
	Triassic	Upper			Jiuligang		T_3j	142	yellowish-grey and dark grey siltstone, mainly sandy shale and mudstone with feldspar quartz sandstone and carbonaceous shale, with 3–7 layers of coal beds or coal lines
		Middle			Badong		T_2b	75–91	purplish-red siltstone and mudstone with grayish-green shale
		Lower			Jialingjiang		T_1j	728	gray medium–thick-bedded dolomite and dolomitic limestone with limestone and evaporite–solution breccia
					Daye		T_1d	1000	gray and light gray thin-bedded limestone with thick-bedded limestone and dolomitic limestone in the middle upper part, and argillaceous limestone or yellowish-green shale in the lower part

Table 2–1

Chronostratigraphic Unit				Rock–Stratigraphic Unit			Code	Thickness/m	Lithological Description
Erathem	System	Series	Stage	Group	Formation	Member			
Upper Paleozoic	Permian	Upper			Wuchiaping		P_3w	84–103	gray medium thick–thick-bedded and massive chert-bearing micritic limestone and bioclastic limestone
		Middle			Maokou		P_2m	88.9	gray and light gray thick-bedded–massive chert-bearing bioclastic microcrystal limestone, algal micrite, and bioclastic arenaceous sparite
					Qixia		P_2q	110.2	dark gray and grayish-black thick-bedded chert-bearing bioclastic micritie
					Liangshan		P_2l	3.8–4.2	grayish-white medium thick-bedded fine-grained sandstone, siltstone, mudstone, and coal beds in the lower part; black thin-bedded mudstone interbedded with limestone in the upper part
	Carboniferous	Middle	Dalaian / Huashibanian		Huanglong		C_2h	11.4	gray–light gray and fleshy-red thick-bedded limestone, containing calcareous dolomitic breccias and agglomerates
			Luosuian		Dabu		C_2d	5.1	grayish-white–grayish-black thick-bedded–massive dolomite
	Devonian	Upper	Famennian		Xiejingsi		D_3C_1x	11.66	sandy shale in the upper part with chamosite siderite and coal lines; marl, limestone, or dolomite with shale and oolitic hematite deposit in the lower part
			Frasnian		Huangjiadeng		D_3h	12.8–15	yellowish-green and grayish-green shale, mainly sandy shale and sandstone, sometimes with oolitic hematite deposit
		Middle	Givetian		Yuntaiguan		$D_{2-3}y$	85.9	grayish-white medium–thick-bedded or massive quartzite-like fine-grained sandstone with grayish-green argillaceous sandstone
Lower Paleozoic	Silurian	Llandovery	Telychian		Shamao	4th	$S_{1-2}sh^4$	51.1–77.4	grayish-yellow and grayish-brown medium–thin-bedded fine-grained sandstone with purplish-red thin-bedded siltstone
						3rd	$S_{1-2}sh^3$	125.5	yellowish-green medium thick-bedded feldspar quartz sandstone with silty mudstone and thin-bedded argillaceous siltstone
			Aeronian			2nd	$S_{1-2}sh^2$	282	yellowish-green thin-bedded silty argillaceous rocks and argillaceous siltstone with grayish-white thin-bedded fine-grained sandstone
						1st	$S_{1-2}sh^1$	185. 3	grayish-yellow and yellowish-green thin-bedded mudstone, grey thin-bedded siltstone, and yellowish-green silty mudstone

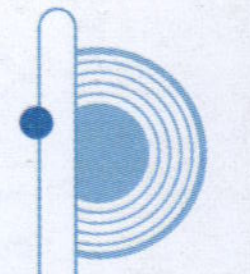

Table 2–1

Chronostratigraphic Unit				Rock–Stratigraphic Unit			Code	Thickness/m	Lithological Description
Erathem	System	Series	Stage	Group	Formation	Member			
Lower Paleozoic	Silurian	Llandovery	Aeronian		Luoreping		S_1lr	73.7–172	yellowish-green mudstone and shale with bioclastic limestone and marl in the lower part; yellowish-green mudstone and silty mudstone in the upper part
					Xintan		S_1x	670–820	grayish-green and yellowish-green shale, sandy shale, and siltstone with thin-bedded fine sandstone
			Rhuddanian		Longmaxi		S_1l	576.5	black and grayish-green thin-bedded silty mudstone and quartz siltstone with thin-bedded quartz fine sandstone, containing lots of graptolites
	Ordovician	Upper	Hirnantian		Wufeng	Guanyinqiao	O_3w^g	0.17–0.3	blackish-gray, yellowish-brown, or light purplish-gray quartz-bearing silty claystone and clay rocks, containing Hirnantia fauna
			Katian			Graptolite Shale	O_3w^s	5.44	blackish-gray micro-thin–thin-bedded organic-containing quartz silty hydromica claystone with blackish-gray micro-thin–thin-bedded microcrystal silicalite
					Linxiang		O_3l		gray, grayish-black, or greenish nodular argillaceous limestone with a few shales
			Sandbian		Baota		O_3b		gray, light purplish-red, or dark purplish-red medium thick-bedded cracked limestone with nodular limestone, containing cephalopods (*Sinoceras sinensis*, etc.)
					Miaopo		$O_{2-3}m$	3.1–6.6	yellowish-green and grayish–black calcareous mudstone, silty mudstone, yellowish-green shale with thin-bedded bioclastic limestone, rich in graptolites

Table 2–1

Chronostratigraphic Unit				Rock–Stratigraphic Unit			Code	Thickness/m	Lithological Description
Erathem	System	Series	Stage	Group	Formation	Member			
Lower Paleozoic	Ordovician	Middle	Darriwilian		Guniutan		O_2g	20.06	cinerous, grey, and purplish-grey thin–medium thick-bedded limestone and conglomerate limestone interbedded with nodular limestone
			Dapingian		Dawan	3rd	$O_{1-2}d^3$	21.55	yellowish-green thin-bedded silty mudstone interbedded with bioclastic limestone in different thicknesses
						2nd	$O_{1-2}d^2$	7.7	purplish-red, grayish-green, or light gray thin-bedded bioclastic micrite and nodular limestone with calcareous mudstone
		Lower	Floian			1st	$O_{1-2}d^1$	25.5	grayish-green, dark gray, and light gray thin-bedded limestone with very thin-bedded yellowish-green shale
					Honghuayuan		O_1h	45.9	gray and dark gray medium–thick-bedded limestone with shale in the lower part occasionally
					Fenxiang		O_1f	22–54	gray medium–thick-bedded limestone with grayish-green thin-bedded mudstone in the lower part; gray thin-bedded bioclastic limestone with mudstone in the upper part
			Tremadocian		Nanjinguan		O_1n	209.77	dolomite in the lower part; chert-bearing limestone, oolitic limestone, and bioclastic limestone in the middle part, containing trilobites; bioclastic limestone with yellowish-green shale in the upper part, rich in trilobites, brachiopods, etc.
					Loushanguan		$Є_3O_1l$	673.37	gray and light gray thin-bedded–massive microfine-grained crystalline dolomite and argillaceous dolomite with breccia dolomite, with chert locally
	Cambrian	Epoch 3	Taijiangian		Qinjiamiao		$Є_3q$		mainly thin-bedded dolomite and thin-bedded argillaceous dolomite with medium–thick-bedded dolomite and a small amount of shale and quartz sandstone

Table 2–1

Chronostratigraphic Unit				Rock–Stratigraphic Unit			Code	Thickness/m	Lithological Description
Erathem	System	Series	Stage	Group	Formation	Member			
Lower Paleozoic	Cambrian	Epoch 2	Duyunian		Shilongdong		$\epsilon_2 sl$	86.3	light gray and dark gray–brownish-gray medium–thick-bedded dolomite and massive dolomite, calcareous and chert-bearing stratum in the upper part
					Tianheban		$\epsilon_2 t$	81–377	dark gray and gray thin-bedded argillaceous banded limestone, rich in fossils of archaeocyatha and trilobites
					Shipai		$\epsilon_2 sh$	294	greyish-green and yellowish-green claystone, sandy shale, fine sandstone, and siltstone with thin-bedded limestone and bioclastic limestone
			Nangaoian		Shuijingtuo		$\epsilon_2 s$	168.5	grayish-black or black shale and carbonaceous shale with grayish-black thin-bedded limestone
		Terreneuvian	Meishucunian		Yanjiahe		$\epsilon_1 y$	20–50	gray siliceous mudstone, dolomite, and black carbonaceous limestone with carbonaceous shale
Neoproterozoic	Sinian	Upper			Dengying	Baimatuo	Z_2dy^b	17.5	grayish-white and thick–medium-bedded dolomite, with siliceous bands and nodules developed in parts of layers
						Shibantan	Z_2dy^s	36	grayish-black thin-bedded siliceous micritic limestone, extremely thin-bedded strips of micritic dolomite developed
						Hamajing	Z_2dy^h	133.4	gray light gray medium-bedded dolomite with thick-bedded dolomite
		Lower			Doushantuo	4th	Z_1d^4	0–8.4	black thin-bedded siliceous mudstone and carbonaceous mudstone with lenticular limestone
						3rd	Z_1d^3	35.8	grayish–white thick-bedded dolomite with medium-bedded dolomite in the lower part; thin-bedded silty dolomite in the upper part
						2nd	Z_1d^2	235	dark grey–black thin-bedded argillaceous limestone and dolomite, interbedded with thin-bedded carbonaceous mudstone in different thicknesses
						1st	Z_1d^1	3.3–5.5	gray and dark grayish-black thick-bedded siliceous-bearing dolomite, with tepee structure developed

Table 2–1

Chronostratigraphic Unit				Rock–Stratigraphic Unit			Code	Thickness/m	Lithological Description
Erathem	System	Series	Stage	Group	Formation	Member			
Neoproterozoic	Nanhuaian				Nantuo		Nh_2n	36–63	grayish-green and purplish-red massive glacial conglomerate, glaical- and gravel-bearing mudstone, and partially thin-bedded silty mudstone
					Liantuo	2nd	Nh_1l^2	39–63	purplish-red and grayish-white tuffaceous sandstone, and purplish-brown and yellowish-green sandstone and sandy shale
						1st	Nh_1l^1	91–103	red, brownish-purple, and yellowish-green coarse–medium-grained feldspar quartz sandstone and feldspar sandstone
Mesoproterozoic				the Kongling Rock Group	Miaowan		Pt_2m	864.12	plagioclase hornblende schist with banded and stripe structures, with quartzite, hornblende plagioclase gneiss, and garnet hornblende schist
					Xiaoyucun		Pt_2x	799.85	graphite-bearing biotite plagioclase gneiss, marble, and calcium silicate rock-quartzite association in the middle and lower parts; amphibolite with biotite plagioclase gneiss, quartz schist, Al-rich gneiss, and schist in the upper part; marble lens occasionally seen at the top
					Gupingcun		Pt_2g	>812	biotite (hornblende) plagioclase gneiss (or granulite) with plagioclase amphibolite

"the Kongling schists", a name given by Li Siguang et al. (1924) at the Three Gorges area of the Yangtze River, Hubei Province. Xie Jiarong et al. (1925) named the gneisses and schists other than Huangling granites in "the Sandouping Group" as "pre-Sinian crystalline schists and gneisses". In Beijing College of Geology (1950), the metamorphic rock series other than the Huangling granites at the core of the Huangling Dome, western Yichang were named as "the Kongling Group". Formed in pre-Sinian, it consists of the Gucunping Formation, the Xiaoyucun Formation, and the Miaowan Formation from bottom to top. Hubei District Survey Team (1984) divided the Kongling Group at the Bangzichang area of Yichang (south to the Huangling granite at the Huangling Dome) into upper, middle, and lower formations, all of which were formed during the Proterozoic. In the Regional Geology of the Hubei Province (1990), "the Kongling Group" referred to all metamorphic rock series with medium–high grades at the core of the Huangling Dome except the Huangling granite and the Taipingxi ultrabasic rock. Within Kongling, the group was further divided into upper, middle and lower formations, belonging to the Neoarchean–Paleoproterozoic. In Exi Geological Team (1990), the geological entity of the Kongling Group referred only to the old metamorphic rock series of the crystalline basement at the south of the Huangling Dome, south to the Huangling granite. Within the group, it was also divided into the Gupingcun Formation, the Xiaoyucun Formation, and the Miaowan Formation from bottom to top, and all formations belonged to the Mesoproterozoic. In 1996, the group was renamed as the Kongling Rock Group in the action of standardizing rock names of Hubei, but the names of the formations in the group are retained as the Gucunping Formation, the Xiaoyucun Formation, and the Miaowan Formation repectively.

1) The Gucunping Formation (Pt_2g)

The Gucunping Formation is a set of metamorphic rock series composed of very thick biotite (amphibole/hornblende) plagioclase gneisses (or granulites) interbedded with plagioclase amphibolites. The assemblage of rocks in this formation is stable and consistent. There are no graphites and marbles in the middle and lower parts, and incidental occurrences of graphite (sillimanites)-bearing biotite plagioclase gneisses in the upper part, which is in conformable contact with the overlying Xiaoyucun Formation, where there are rich graphite-bearing gneisses. The lower part is imcomplete due to the intrusion of Huangling granites. The thickness of the Gucunping Formation is greater than 812 m.

Geological characteristics and regional variations: The strata of this formation are outcropped at places of Tiaoyutan, Hongguixiang, Meijawan, and Changling–Shipailing in Gucunping, Dengcun, Yichang. They constitute the northern wing of the Meishichang syncline. From the bottom up, the plagioclase amphibolite interbeds decrease, with only a small amount of biotite arkosites, graphite (sillimanite)-bearing biotite plagioclase gneisses, and biotite granulites. Generally speaking, the extension of the formation is stable and rocks in

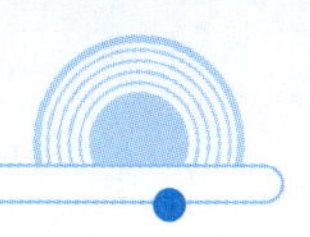

the formation are unified with mild migmatization. According to the geochemical characteristics, the original rocks of this formation were basaltic, anglicic, andesitic, and rhyolitic volcanic rocks, with few terrigenous clastic rocks.

2) The Xiaoyucun Formation (Pt_2x)

The middle and lower parts of the Xiaoyucun Formation are graphite-bearing biotite plagioclase gneiss, marble, and calcium silicate-quartzite assemblages. The upper part is composed of plagioclase amphibolites intercalated with biotite plagioclase gneisses, quartz schists, Al-rich gneisses, and schists. There are also occasional presences of marble lenticles at the top. The bottom part is marked by the presence of abundant graphite-bearing gneisses and arkosites, distinguishing the Xiaoyucun Formation with the underlying Gucunping Formation. The two formations are in conformable contact. The upper part of the formation is in comformable contact with the Miaowan Formation. The thickness of the Xiaoyucun Formation is 799.85 m.

Geological characteristics and regional variations: The strata of this formation are mainly outcropped at places such as Tianbaoshan, Xiaoqucun, Houzizhai, Qinglongbao, Guojiaya, Xiaoxikou, Baihubao, Beipingya, and Chengshuping in Meizhichang, Yichang. They constitute the Meishichang syncline, the Duanfangxi anticline, and wings of some small-scaled NE-trending folds. The basal of the formation is marked with a combination of arkosites and abundant graphite-bearing gneisses. It is bordered with the underlying Gucunping Formation in conformable contact. The lower part is composed primarily of biotite plagioclase gneisses intercalated with graphite-bearing biotite plagioclase gneisses, graphite-bearing biotite schists, biotite arkosites, garnet-bearing biotite plagioclase gneisses, garnet-bearing andalusite mica-quartz schists, biotite sillimanites, and alusite schists, and a small amount of biotite granulites and amphibolites. In other words, it is characterized by gneisses rich in quartzites, graphites, and Al-rich minerals, which constitute the Al-rich lower part of the formation. The middle part is mainly composed of marbles and calcium silicates, intercalated with biotite plagioclase gneisses and a small amount of quartzites and amphibolites. In this part, the calcite dolomite (or dolomite calcite) marbles coexist with calcium silicates such as bistagites, tremolite-bearing bistagites, calcite-bearing plagio-bistagites, and calcite-bearing bistagites. And this part stretches steadily in area, constituting a salient feature. The upper part is mainly composed of plagioclase amphibolites, intercalated occasionally with various amphibolites, quartzs, and amphiboles or quartz schists rich in felsic, calc-silicates, ferromagnesian, and aluminous elements. Besides, there are also outcrops of various quartzites. All those constitute a unique formation of this part. At the same time, garnets are pervasively seen in various rocks in the upper part and often coexist with andalusites, sillimanites, kyanites, and corundums, which contribute to an Al-rich layer of the upper part of the Xiaoyucun Formation.

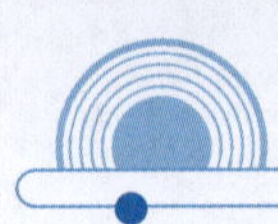

3) The Miaowan Formation ($Pt_{2-3}m$)

The Miaowan Formation is composed of a set of very thick-bedded and unitarily featured plagioclase hornblende schists with banded or striped structures, intercalated with quartzites, hornblende plagioclase gneisses, and garnet amphibolite schists. The presence of very thick- and thick-bedded plagioclase hornblende schists separates the formation from the underlying Xiaoyucun Formation. The two formations are in conformable contact. The top of the formation is unconformably covered by the Liantuo Formation. The thickness of this formation is 864.12 m.

Geological characteristics and regional variations: The strata of this formation are outcropped at places of Quejiaping, Miaowan, Qingshuling, and Huanxiya in Meizhichang, Yichang. They constitute the core of the Meizhichang syncline. The formation is mainly composed of thin-bedded, medium–thick-bedded, and very thick-bedded plagioclase hornblende schists, which are developed into silica bands and striped structures, intercalated with quartzites and horblende plagioclase gneisses. The formation stretches steadily in the area. The petro-geochemical characteristics of the plagioclase hornblende schists reveal that they were formed through the metamorphisim of basalts from marine eruption. Therefore, this formation represents the production of basaltic magma eruption at the late forming stage of the Kongling Rock Group.

4) About the age of the Kongling Rock Group

The original age for the formaction of the Kongling Rock Group is presumed to be the Mesoproterozoic. And evidences for such a presumption are as follows:

(1) The Kongling Rock Group is not only invaded by the Huangling granites and the Taipingxi ultrabasic rocks, but also imcompletely covered by the Liantuo Formation.

(2) Exi Geological Team once collected abundant acritarch fossils in the Xiaoyucun and Miaowan Formations of the Kongling Rock Group. Some of the acritarch fossils were with thin shells and simple ornaments similar to those in Changchengian of North China, and were still relatively primitive members such as *Trachysphaeridium simplx* and *Leiopsophosphaera minor*. "The distributions of these fossils are relatively stable. Although nearly 20 years has passed since the establishment of these genera and species, they have not yet been found in higher strata" (see Xing Yusheng's *Late Precambrian Palaeontology in China*). However, most of them are in Changchengian–Jixianian that could connect the previous with the latter or diachronous, for example, *Leiopsophosphaera apertus*, *L. densa*, *Asperatopsophosphaera umishanensis*, etc. On the whole, the acritarch fossils from the above the Kongling Rock Group constituted the micropalaeoflora in Changchengian–Jixianian.

(3) Exi Geological Team collected four groups of zircons from the graphite-bearing biotite gneisses in the Xiaoyucun Formation, the Kongling Rock Group for Pb-Th isotope age determination. The obtained data indicate that the age value on the higher crossing point

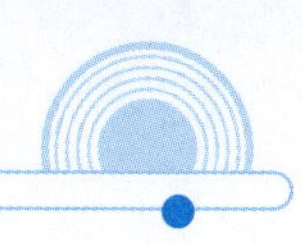

of Wetherill concordance diagram is (1991 ± 30) Ma. At the same time, six groups of plagioclase amphibolite samples were taken from the Miaowan Formation for Sm-Nd isochron determination. And the results revealed that the whole-rock isochron age was (1608±81) Ma.

(4) The Sm-Nd isochron age is (1282 ± 86) Ma for the various basic–ultrabasic rock units in Meizhichang—intrusions into Taipingxi basic–ultrabasic complex of the Kongling Rock Group. The age of the Sandouping hornblende biotite palites that intruded into the Kongling Rock Group is (931 ± 38) Ma, according to the ziron U-Pb isotopic dating data shown on the crossing point of the Wetherill concordance diagram. Besides, the Huangling granite series that constitute the main component of the Huangling granite complex not only encroached into the Kongling Rock Group, but also the Meizhichang and Maoping rock series, including the Sandouping intrusions. These intrusives are incompletely covered by the Nanhuaian Liantuo Formation. Therefore, their age and mutual relations are important factors in determining whether the formation age of original rocks belongs to the Mesoproterozoic.

2. The Neoproterozoic

The Neoproterozoic at the fieldwork area includes the Qingbaikou (Tonian) System the Nanhuaian (Cryogenian) System and the Sinian (Ediacaran) System. From bottom to top, it is divided into the Liantuo Formation, the Nantuo Formation, the Toushantuo Formation, and the Dengying Formation.

1) The Liantuo Formation ($Nh_1 l$)

The Liantuo Formation is evolved from "the Liantuo Group", which was established by Liu Hongyun and Sha Qingan (1963). The formation was once put under the lower part of the Nantuo Formation by Li Siguang et al. (1924), Wang Rilun (1960), Zhao Zongpu (1954), Hubei Institute of Geological Science (1962), Hubei Regional Surveying and Mapping Team (1970), and Central South Stratigraphic Chart Abbreviation Team (1974). Liu Hongyun et al. (1963) separated the formation from the Nantuo Formation and renamed it as the Liantuo Group. Later, Three Gorges Strata Research Group (1978) changed it into the Liantuo Formation and this name has been widely used ever since.

The Liantuo Formation refers to the set of aubergine to claret medium–thick-bedded glutenites, gravel-bearing coarse-grained sandstones, feldspathic quartz sandstones, quartz sandstones, fine lithic sandstones, and feldspathic sandstones intercalated with tuffaceous lithic sandstones, and gravel-bearing lithic tuffs, that exist between the Huangling granites and the Nantuo Formation. From bottom to top, the grain size of the debris in this formation varies from coarse to fine. The top of the formation is in parallelly unconformable contact with the glacial conglomerate at the base of the Nantuo Formation. The bottom of the Liantuo Formation is in unconformable contact with the Huangling granites. The lithology of this formation can be divided into two members: the Lower Member are aubergine or brown medium–thick-bedded glutenites, gravel–bearing coarse sandstones, feldspathic sandstones, tuffaceous sandstones, tuffs, etc. Conglomerates might sometimes be seen at the base. The thickness of the Lower Memeber is 39–63 m. The Upper Member are aubergine or grayish–green medium–thick-bedded fine-grained lithic sandstones, feldspathic sandstones intercalated with tuffaceous lithic sandstones, crystal pyroclasts and crystal-vitric tuffs, with a thickness of 91–105 m.

According to Zhao Ziqiang et al. (1985), this formation contains 11 genera and 19 species of micropalaeoflora, most of which belong to the subgroup of Chlorella, such as *Leiopsophosphaera minor*, *Trachysphaera plamum*, etc. In addition, the U-Pb age of zircon collected by Zhao Ziqiang et al. (1985) from tuffs in the Liantuo Formation of East Three Gorges is (748±12) Ma.

2) The Nantuo Formation (Nh_2n)

The formation was first named by Blackwelder (1907) according to the place of Nantuo, Yichang. Li Siguang et al. (1924) defined it as the Nantuo Beds and Wang Rilun (1960) defined it as the Nantuo Formation. Later, Liu Hongyun and Zhao Qingan (1963) revised and redefined the Nantuo Formation as the Nantuo Beds in Li et al. (1924) or the tillites in the Nantuo Formation of Wang (1960). Thereafter, the name of the Nantou Formation has been widely used.

The Nantuo Formation is composed of grayish-green and aubergine glacial conglomerates (anagenites). The upper part of the formation is intercalated with layered sandstone lenticles. The gravels in the glacial conglomerate (anagenites) are of poor sorting and have scratches at the surface. The formation is in parrellel but unconformable contact with the dolomites in the overlying Doushantuo Formation, and the tuffaceous fine sandstones in the underlying Liantuo Formation. The thickness of this formation is 50–200 m.

Zhao Ziqiang et al. (1988) found 24 species across 11 genera in the Nantuo Formation, including *Leiopsophosphaera minor* and *Trachysphaeridium rugosum* of the sphaeropsis subgroup and banded algae and brown algae fragments of the cylindrical algae subgroups.

3) The Doushantuo Formation (Z_1d)

The Doushantuo Formation is evolved from "the Doushantuo Rock Series" created by Li Siguang et al. (1924). The place they referred to is located at Doushantuo, Yichang. Beijing College of Geology (1961) put this section of strata under the lower part of the Dengying Group, and named it as the Doushantuo Formation. The Institute of Geology in Chinese Academy of Geological Sciences (1988) put it under the lower part of the Dengying Formation and named it as the Doushantuo Beds. Liu Hongyun et al. (1963) changed the name into the Doushantuo Formation, and the name has been used ever since.

The Doushantuo Formation underlies conformably under the Dengying Formation and

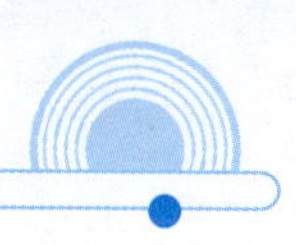

overlies in parallelly unconformable contact with the Nantuo Formation. The top of the formation is demarcated with the overlying Dengying Formation by black carbonaceous shales. The bottom is demarcated with the underlying Nantuo Formation by a layer of cap dolomites. From bottom to top, the formation is divided into four members. The First Member is composed of gray, dark gray and black thick-bedded dolomites with siliceous- and chert-nodules, thin–medium-bedded dolomites and limy dolomites, with a thickness of 3.3–5.5 m. The Second Member is composed of dark gray–black thin-bedded argillaceous limestones and dolomites intercalated with thin-bedded carbonaceous mudstones. The layers are interbedded with unequal thickness and rhythm, containing argillaceous and siliceous phosphorus nodules. The thickness of the member is 235 m. The lower part of the Third Member is composed of incanus and thick-bedded dolomites interbedded with medium–thick-bedded dolomites and silty–fine-grained dolomites, containing chert nodules and stripes developed. The upper part of the Third Member is composed of thin-bedded silty dolomites. The thickness of the member is 835 m. The Fourth Member is composed of black thin-bedded siliceous mudstones and carbonaceous mudstones intercalated with lenticular limestones. The thickness of the member is 0–8.4 m.

There are abundant micropalaeophyte fossils in the black shales and phosphorous dolomites of the Doushantuo Formation. According to Zhao Ziqiang et al. (1988), there are about 90 species across about 50 genera, including mainly the Aegagropila subgroup, the Nitzschia subgroup, and the open spherical microflora. In particular, the occurance of calcareous sponges, siliceous sponges, and chitinozoa in the formation indicates that this formation belongs to the early Sinian.

4) The Dengying Formation (Z_2dy)

The Dengying Formation is evolved from "the Dengying limestones" set up by Li Siguang et al. (1924). The naming place is located at Dengyingxia, which is between Shipai Villiage and Nantuo Villiage. The two villiages are at the south of the Yangtze River, 20 km away from the northwest of Yichang. The strata here were once named as "the Dengying Formation of the Upper Dengying Group" by Beijing College of Geology (1961). The Academy of Geological Sciences (1962) named the Doushantuo Beds and the Dengying limestones collectively as the Dengying Formation. Liu Hongyun et al. (1963) named the Dengying limestones as the Dengying Formation, which was adopted by others later. Zhao Ziqiang et al. (1985) once divided the Dengying Formation into the Hamajing Member, the Shibantan Member, the Baimatao Member, and the Tianzhushan Member from bottom to top. Nowadays, the Tianzhushan Member is recognized as belonging to the Cambrian, so the Dengying Formation contains only three members.

The Dengying Formation refers to a set of strata which is in parallel unconformity with the underlying Niutitang Formation (the Shuijingtuo Formation) and in conformity with the top of the overlying Doushantuo Formation. The lithology in this formation can be divided into three members. ①The lower Hamajing Member, with a thickness up to 134.4 m, consists mainly of intraclastic dolomites and fine-grained dolites associated with minor silicon-bearing fine-grained dolites, which are gray–light gray and show medium–thick-bedded formation. ②The middle Shibantan Member, with a thickness up to 36 m, consists of dark gray and grayish-black thin-bedded silicon-bearing micritic limestones intercalated occasionally with chert bands and development of very thin-bedded micritic dolomite bands. The member contains macro algae. ③The upper Baimatuo Member, with a thickness up to 17.5 m, consists of offwhite thick–medium-bedded dolomites interbedded with medium–thin-bedded fine-grained dolomites, with siliceous bands and nodules development in some places of the member. The siliceous phosphorous dolomites at the top of the member produce small shelly fossils.

There are systematic collection and research on the paleontology of the Dengying Formation at the East Three Gorges area. The studies demonstrate that there are 55 species across 25 genera, as well as metaphytes such as *Vendotaenia* and *Tyrasotaenia*, metazoa mollusk, and their tracefossils. The top of the formation produces small shelly fossils and the formation was formed during the late Sinian.

2.1.2.2 *The Palaeozoic*

In the fieldwork area, the Lower Cambrian includes the Yanjiahe Formation (the Tianzhushan Member), the Shuijingtuo Formation, the Shipai Formation, the Tianheban Formation, the Shilongdong Formation, the Qinjiamiao Formation, and the Cambrian–Ordovician Loushanguan Formation. The Ordovician includes the Nanjinguan Formation, the Fenxiang Formation, the Hunghuayuan Formation, the Dawan Formation, the Guniutan Formation, the Miaopo Formation, the Baota Formation, and the Wufeng Formation. The Silurian includes the Longmaxi Formation, the Luoreping Formation, and the Shamao Formation.

1. The Cambrian System

1) The Yanjiahe Formation ($Є_1y$)

The Yanjiahe Formation was named by Ma Guogan and Chen Guoping (1981). The accurate place of the formation is located at Yanjiahe of Sandouping, Yichang. It originally belonged to the non-trilobite member at the base of the Shuijingtuo Formation. The formation, with a thickness of 20–50 m, consists mainly of gray siliceous mudstones, dolomites, and black carbonaceous limestones intercalated with carbonaceous shales. It, together with the underlying Dengying Formation and the overlying Shuijingtuo Formation, belongs to comformable deposition. The small shelly fossils can be divided into the upper and lower assemblies, with the lower assembly including *Circotheca-Anabarites-Protohertzina* and the

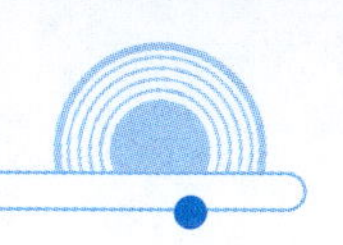

upper assembly including *Lophotheca-Aldanella-Maidipingoconus*. They can be compared with the first and second assemblies of the Meishucun Stage.

2) The Shuijingtuo Formation (Є$_2$*s*)

In 1957, Zhang Wentang discovered a new set of trilobite fauna layer from the lower Shipai shale section established by Li Siguang (1921). In the layer, no *Redlichia* was found. Thus, Zhang named the layer as the Shuijintuo Formation. The standard stratotype section of this formation is at Shuijingtuo, about 400 m southeast of Shipai, Yichang. After being revised by Beijing Collgeg of Geology (1960) and Zhang Shusen et al. (1978), the Shuijingtuo Formation refers to a formation consisting of grayish-black or black shales, and carbonaceous shales intercalated with black thin-bedded limestones. The formation contains trilobites including *Sinodiscus shipaiensis*, *S. similis*, *S. changyangensis*, *Tsunyidiscus ziguiensis*, *T. sanxiaensis*, *Hupeidiscus orientalis*, and *H. elevatus*, as well as brachiopods, spongy spicules, hyolithes, etc. The formation is in conformity or parallel unconformity with the underlying stratum.

The lithology of the Shuijingtuo Formation is relatively stable. Usually, the lower part consists mainly of carbonaceous shales intercalated with limestones or dolomitic limestones. The upper part consists mainly of black limestones or carbonaceous shales. The formation is of 168.5 m thick. The trilobites of the formation are mainly *Paigetia* of the subfamily of *Eodiscina*. The trilobites in this formation are named the Xiadong Fauna. Besides, the formation also contains members of Redlichiidae in Redlichiida. They are all typical faunal members of Early Cambrian in Hubei. It can be compared with the Guojiaba Formation (or the Shuijingtuo Formation) in Nantan, Ningqiang, Mianmu, Zhenba, and other places of Shaanxi, the Qiongzhusi Formation in the southwestern of Sichuan and Yunnan, and the Liangshuijing Formation in Chengkou, Sichuan. The formation belongs to the eastern Guizhou system.

3) The Shipai Formation (Є$_2$*sh*)

The Shipai Formation is evolved from "the Shipai shales" created by Li Siguang et al. (1924). The formation is located at Shipai Villiage, which is on the south bank of Yangtze River, 20 km north to the Yichang City. After rounds of revision by Zhang Wentang et al. (1957), Beijing College of Geology (1960), and Hubei Three Gorges Strata Team (1978), "Shipai shales" has been renamed as the Shipai Formation, which refers to the non-black rock series overlying the black strata in the east area of Three Gorges, western Hubei, a definition given by Hubei Three Gorges Strata Team (1978).

The Shipai Formation, containing trilobite fossils, is composed of a set of grayish-green to yellowish-green claystones, sandy shales, fine sandstones, siltstones interbedded with thin-bedded limestones, and bioclastic limestones. At the base of the formation, the grayish-green sandy shales are in conformable contact with the black shales intercalated with grayish-black thin-bedded limestones in the Shuijingtuo Formation. At the top boundary, the shales and limestone-bearing siltstones are in conformable contact with the gray argillaceous-banded limestones in the Tianheban Formation.

The Shipai Formation is abundant in fossils, in particular the trilobites of *Redlichia*. The main trilobites in the formation are *R. kobayashi*, *R. meitanensis*, *Palaeolenus lantenoisi*, *Kootenia yichangensis*, *Ichangia conica*, *Neocobboldia hubeiensis*, etc. There are also some brachiopods, which indicates that this formation should be formed during the early Duyunian.

4) The Tianheban Formation (€$_2$*t*)

The Tianheban Formation is evolved from "the Tianheban limestones", established by Zhang Wentang et al. (1957). The formation is located at Tianheban, which is between Shipaicun Village and Shilongdong Cave, 20 km away from the northwest of Yichang.

The Tianheban Formation is in conformable contact with the underlying Shipai Formation and the overlying Shilongdong Formation. It is mainly composed of dark gray–gray thin-bedded argillaceous banded limestones, intercalated occassionally with a few yellowish-green shales and oolitic limestones in some places. The formation contains abundant archaeocyathid and trilobite fossils. The argillaceous-banded limestones at the base of the formation demarcate it from the grayish-green thin-bedded sandy shales in the Shipai Formation. At the top, the argillaceous-banded limestones demarcate the formation from the thick-bedded dolomites in the Shilongdong Formation. The thickness of this formation is about 88–108 m.

The Tianheban Formation contains abundant archaeocyathid and trilobite fossils, with the former including *Archaeocuthus hupeiensis*, *A. yichangensis*, *Retecyathus kusmini*, *Protopharetra* sp., *Sanxiacyathus hubeiensis*, *S. typeus*, etc., and the latter including *Megapalaeolenus deprati*, *M. obsoletus*, *Palaeolenus minor*, *Kootenia Ziguinensis*, *Xilingxia convexa*, *X. yichangensis*, etc. The formation is formed during the middle Duyunian.

5) The Shilongdong Formation (€$_2$*sl*)

The Shilongdong Formation is evolved from "the Shilongdong limestones" created by Wang Yu (1938). The place is located at Shilongdong Cave, which is at the south bank of the Yangtze River, 18 km away from the northwest of Yichang. Later on, Zhang Wentang et al. (1967) redefined "the Shilongdong limestones" (narrow meaning) as the large set of thick-bedded dolomites without archaeocyathids in the middle and upper parts of the original "the Shilongdong limestones" (broad meaning). Since then, the definition has been widely accepted and renamed as the Shilongdong Formation.

The Shilongdong Formation, with a thickness of up to 86.3 m, is composed mainly of a set of light gray–dark gray medium–thick-bedded dolomites and massive dolomites. The top of the formation contains a small amount of calcareous and chert clusters. The thick-bedded dolomites at the base of the formation are in conformable contact with the argillaceous-banded limestones in the underlying Tianheban Formation, and the thick-bedded dolomites at the top are in conformable contact with the overlying Qinjiamiao Formation.

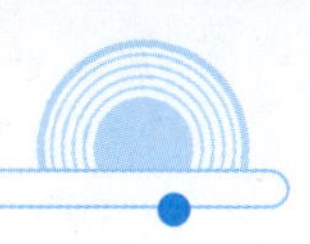

This formation contains few fosssils. Only tribobite fossils were found at Zhenzhukou of Tongshan, Jiangsu, and Zhujiayu of Nanzhang, Hubei. At Zhenzhukou, *Relidlichia* sp. and *Yukesinzella* sp. were collected. At Zhujialing, *Redlichia* sp., *R.* (*Redlichia*) *guizhouensis coniformis*, and *R.* (*Pteroredlichia*) *murakami* were collected. According to those trilobites identified, the Shilongdong Formation should be formed during the late Duyunian.

6) The Qinjiamiao Formation ($€_3q$)

The Qinjiamiao Formation is evolved from "the Qinjiamiao thin-bedded limestones" created by Wang Yu (1938). The place is located at Qinjiamiao, Yichang. Lu Yanhao (1968) changed "the Qinjiamiao thin-bedded limestones" into "the Qinjiamiao Group". Sun Zhenhua (1988) named it "the Qinjiamiao Formation". Hubei Regional Surveying and Mapping Team (1968) called it "the Maoping Formation". Wang Xiaofeng et al. (1987) restored its name as "the Qinjiamiao Group". At last, the name was settled on "the Qinjiamiao Formation" by Hubei Bureau of Geology and Mineral Resources (1996) after discussing and seeking agreement with neighboring provinces and regions.

The Qinjiamiao Formation refers to a set of strata inbetween the two sets of carbonates in the Shilongdong Formation and the Loushanguan Formation (formerly the Sanyoudong Formation) respectively. The formation consists mainly of thin-bedded dolomites and thin-bedded argillaceous dolomites, intercalated with medium–thick-bedded dolomites and a small amount of shales and quartz sandstones. There are often ripple marks, mud crack structures, and halite and gypsum pseudocrystals in this rock formation. Its top is in conformable contact with medium–thick-bedded dolomites in the Loushanguan Formation. Its bottom is in conformable contact with thick-bedded dolomites in the Shilongdong Formation.

The main members of trilobites acquired from this formation are *Solenoparina trogus*, *S.*? *pingshanpaensis*, *Xingrenasipis grenaspts*, *Schopfaspis hubeiensis*, *S. zhaojipingensis*. There are also brachiopod fossils found in this formation. Accordingly, the Qinjiamiao Formation should be formed during Taijiangian.

7) The Loushanguan Formation ($€_3O_1l$)

The Loushanguan Formation is developed from "the Loushanguan limestones" created by Ding Wenjiang (1930) and published in 1942. The founding place is located at Loushanguan, a place inbetween Zunyi and Tongzi in Guizhou. Ever since then, the name has been widely used by many geologists.

The Loushanguan Formation, lying between the Nanjingguan Formation and the Qinjiamiao (or Shilongdong) Formation, is a set of stratigraphic sequences consisting mainly of gray–light gray thin-bedded–massive microcrystalline dolomites and argillaceous dolomites associated with breccial dolomite. There are also cherts in some places of the formation. The base of the formation is demarcated from the Qinjiamiao Formation with the disappearance of gray and grayish–green argillaceous dolomitic siltstones and medium thick- and thick-bedded dolomites. The disappearance of light gray and grayish-white medium–thin-bedded dolomites at the top of the formation demarcates it from the bioclastic limestones in the Nanjinguan Formation. The thickness of the formation is 673.37 m.

There are few fossils in this formation. At present, trilobite fossils have been obtained from some high-calcite-bearing limy dolomites, dolomitic limestones, and limestones in places such as Dingzhai and Tuleping in Xianfeng, Chunmucao, Chashan to Taiyanghe in Enshi, and Xinping in Yichang. The strata of the lower upper part of the formation contain trilobites such as *Paranomocare hubeiensis*, *P. guizhouensis*, *Paramenocephalites acis*, *Xianfengia binodus*, *X. puteata*, *Crepicephalina hubeiensis*, and *Poshania* sp. Accordingly, strata of this layer should be formed during the late Middle Cambrian. The lower middle part of the formation produces trilobites such as *Fangduia subeylindrlca*, *Artaspis xianfengensis*, *Liaoningaspis sichuanensis*, *Stephanoare* sp., and *Blackwelderia* sp. Therefore, the lower middle part of the formation should be formed during the early Late Cambrian. The trilobites produced in the upper middle section are *Enshia typical*, *E. brevica*, etc., and in the upper part are *Saukia enshiensis*, *Calvinella striata*, together with some conodonts such as *Teridontus nakamurai*, *Eoconodontus notchpeakensis*, and *Cordylodus proavus*. According to those fossils, the middle and upper parts of this formation was formed during the middle–late Late Cambrian. At the top of the formation, that is, 7–40 m from the top boundary, there is occurrence of conodonts, mainly *Hirsutodontus simplex*, *Monocostodus sevierensis*, etc. Therefore, this section should be formed in the early Early Ordovician. To sum up, this formation was formed during Wangcunian (Cambrian) and Tremadocian (Early Ordovician), and it is a lithostratigraphic unit with comparatively long time spans.

2. The Ordovician System

1) The Nanjinguan Formation (O_1n)

The Nanjinguan Formation is evolved from "the Nanjinguan limestones" created by Zhang Wentang (1962). The founding location is at the place of Nanjinguan, Yichang, Hubei. The stratum of this place belongs to the upper section of Li Siguang and Zhao Yaceng's (1924) "the Yichang Limestones", the top section of Wang Yu's (1938) "the Sanyoudong limestones" and the Fenxiang Series, the lower and middle sections of Xu Jie and Ma Zhentu's (1948) "the Yichang formation", Yang Jingzhi and Mu Enzhi's (1951, 1954) "the Yichang formation" and "the Fenxiang shales", or a combination of both the Nanjinguan Formation and the Fenxiang Formation in Zhang Wentang (1962) and Hubei Bureau of Geology and Mineral Resources (1978). Hubei Bureau of Geology and Mineral Resources (1996) combined the Nanjinguan Formation and the Fenxiang Formation together as the Nanjinguan Formation, which is adopted in this book.

The Nanjinguan Formation refers to the stratigraphic sequences that are in conformable contact between the Loushanguan Formation and the Honghuayuan Formation. It consists mainly of carbonate rocks, which are light gray–gray and show medium–thick-bedded formation. The base of the formation is composed of bioclastic limestones and limestones, containing trilobite and brachiopod fossils. The lower section is composed of dolomites. The middle section is composed of chert-bearing limestones, oolitic limestones, and bioclastic limestones, and contains trilobite fossils. The upper section is composed of bioclastic limestones interbedded with yellowish-green shales. The section is abundant in trilobites and brachiopods. The base boundary is marked by the presence of bioclastic limestones. The thickness of the formation is about 209.77 m.

There are abundant trilobite, graptolite, and brachiopod fossils in both the base and upper section of this formation. From bottom to top, there are abundant conodonts, cephalopods, and ostracods. Trilobites include mainly *Asaphellus inflatus*, *Dactylocephalus dactyloldes*, *Asaphopsis immanis*, *Szechuanella szechuanensis*, *Tungtzuella szechuanensi*, etc. Graptolites include mainly *Dictyonema asiaticum*, *D. belliforme yichangensis*, *Callograptus curoithecalis*, *Dendrograptus yini*, *Acanthograptus sinensis*, *Adelograptus* sp., etc. Conodonts include mainly *Codylodus angulodus*, *Sanxiagnathus sanxiaensis*, *Acanthodus costalus*, *Glyptoconus quadraplicatus*, *Paltodus deltifer pristinus*, *P. dehifer deltifer*, *Acodus hamulus*, *Drepanoistodus pitjanti*, *Triangulodus bicostatu*, etc. Those fossils indicate that this formation should be formed during the early ages of Early Ordovician.

2) The Fenxiang Formation ($O_1 f$)

The name of the Fenxiang Formation was derived from Wang Yu's (1938) the Fenxiang Series by Zhang Wentang (1962). The founding location is at the northern hillside of the Nvwamiao, West Fenxiang, Yichang.

The lower section of the Fenxiang Formation consists mainly of gray medium-thick-bedded sandy bioclastics and globulitic limestones asscociated with grayish-green thin-bedded mudstones, which are interbedded in non-uniform thickness. The upper section consists mainly of gray thin-bedded bioclastic limestones associated with mudstones. With a thickness of about 22-54 m, the formation contains a wide range of fossils. Graptotites are mainly distributed at the upper section of the formation, which can be broadly divided into two zones: a lower zone of *Acanthograptus sinensis* and an upper zone of *Adelograptus-Kiaerograptus*. The trilobites in the formation are mainly *Dactylo phallus*, *Psilocephalina*, *Szechuanella*, *Asaphopsis*, *Tunghzuella*, *Goniophrys*, *Coscnia*, *Protopliomerops*, and *Parapilekia*. The conodonts in the formation are mainly *Paltodus deltifer*, *Acodus hamulosus*, and *Paroitodus inistus*. The formation was formed during Tremadocian.

3) The Honghuayuan Formation (O_1h)

The name of the Honghuayuan Formation was originally created by Zhang Mingshao and Sheng Xinfu (1940), and further evolved from Liu Zhiyuan's (1948) "the Honghuayuan limestones". Zhang Wentang (1962) once named it as "the Honghuayuan limestone formation". Later, Mu Enzhi et al. (1979) used "the Honghuayuan Formation" to refer specficially to the grayish-black thick-bedded limestones at the Huanghuachang section, which contains abundant cephalopod, brachiopod, and sponge fossils. From then on, Mu's definition has been widely adopted as regards to the lithologic features of the formation.

The Honghuayuan Formation is in conformable contact with the underlying Nanjinguan Formation (limestones or shale-bearing limestones) and the overlying Dawan Formation (shales). The formation is composed of limestones and bioclastic limestones, which are gray or dark gray and show medium–thick-bedded formation associated with thin-bedded micritic–macrocry and bioclastic limestones. The formation often contains chert nodules and phacoids, with shales occassionaly interbedded in the lower section. It also contains abundant fossils of cephalopods, sponge spicules, trilobites, and brachiopods, with calathium biogenetic reefs in some parts of the formation. The thickness of the formation is 45.9 m.

Besides cephalopod and sponge spicule fossils, the formation contains conodont, brachiopod, and trilobite fossils as well. The cephalopods in this formation include *Coresanoceras*, *Manchuroceres*, *Clitendoceras*, *Oderoeras*, *Chaohuceras*, *Recorooceras*, *Hopeioceras*, *Kerkoceras*, *Teratoceras*, *Belmnoceras*, etc. The sponge spicules include *Achaeocyathus* (*Achaeocyathus*) *chihiensis*, etc. The main conodonts are mainly *Triangulodus bicostatus*, *Tropodus yichangensis*, *Acodus suberectus*, and *Serratognathus* sp. The formation was formed during Early Ordovician.

4) The Dawan Formation ($O_{1-2}d$)

The Dawan Formation is evolved from "the Dawan layer" created by Zhang Wentang et al. (1957). The founding location is at the Dawan of Nvwamiao, Fenxiangchang, Yichang. Zhang Wentang (1962) expanded "the Dawan layer" to include Ji Rongsen's (1940) "the Meitan shales", and Li Siguang's (1924) "the Yangzibei layer" and renamed it as the Dawan Formation, which has been adopted ever since.

The Dawan Formation is a set of carbonate strata with high argillaceous content and rich in brachiopods, trilobites, and graptolites. It is in conformable contact with the medium-thick-bedded limestones in the overlying Guniutan Formation and the dark gray thick-bedded cephalopod-bearing macrocrystalline bioclastic limestone in the underlying Honghuayuan Formation. From bottom to top, the formation could be divided into three sections. The first section, with a thickness up to 25.5 m, consists mainly of bioclastic-bearing micrites and microcrystal limestones associated with yellowish-green very thin-

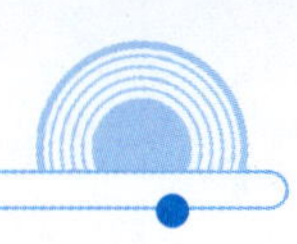

bedded shales, which are grayish-green, dark gray, or light gray and show thin-bedded formation. The second section, with a thickness of about 7.7 m, consists mainly of bioclastic micrites and nodular micrites, associated with few calcareous mudstones, which are purple, grayish-green, or light gray and show thin-bedded formation. The third section, with a thickness of about 21.55 m, consists mainly of silty mudstones associated with bioclastic limestones, which are yellowish-green and show thin-bedded formation, and interbedded with one another in non-uniform thickness.

The formation contains especially abundant fossils, among which, the graptolites, the conodonts, the cephalopods, the brachiopods, and the trilobites have been studied comparatively thoroughly. According to the study of Wang Xiaofeng et al. (2008), the graptotites in the Dawan Formation can be devided into four zones from bottom to top: ①the *Didymograptus bifidus* Zone, ②the *Azygograptus suecicus* Zone, ③the *Isograptus caduceus imitates* Zone, and ④the *Exigraptus clavus* Zone. According to the study of Ni Shizhao et al. (1987), the conodonts in the formation can be divided into four zones as well from the bottom up: ① the *Oepikodus evae* Zone, ② the *Baltoniodus triangularis* Zone, ③the *Baltoniodus navis-Paroistodus parallelus* Zone, and ④ the *Paroistodus originalis* Zone. Among them, Zones 1–3 are located at the lower section of the formation and Zone 4 at the middle section. According to Lai Caigen and Xu Guanghong (1987), the cephalopods in this formation can be devided into three zones from bottom to top: ①the *Bathmoceras* Zone, ②the *Protocycloceras depraty* Zone, and ③the *Protocycloceroides-Cochlioceroides* Assemblage Zone. In terms of brachiopods, the genera of *Yangtzeella*, *Sinorthis*, *Martellia*, *Leptella*, *Lepidorthis*, and *Euorthisina* are particularly prosperous. As for trilobites, 37 genera and species have been identified according to Lu Yanhao (1975). Xiang Liwen and Zhou Tianmei (1987) have further divided them into two zones: the upper *Pseudocalymenea cylindrica* Zone, and the lower *Hanchungolithus* (*Ichangolithus*) Zone. In summary, this formation was mainly formed during Floian of Early Ordovician–Dapingian of Middle Ordovician, with the top section entering in Darriwilian. Among them, the Golden Spikes of the Dapingian are at Huanghuachang, Yichang, with the first appearance of conodont *Baltoniodus triangularis* marking the base boundary of Dapingian.

5) The Guniutan Formation (O_2g)

The Guniutan Formation is evolved from "the Guniutan limestones" created by Zhang Wentang et al. (1957). The founding location of the formation was at Guniutan, Fenxiangchang, Yichang. It refers to the set of strata sequences in conformable contact between the Dawan Formation and the Miaopo Formation. The strata of this formation, rich in cephalopod and trilobite fossils, are composed of interbedded bioclastic micrites, calcirudites, and nodular argillaceous limestones, which are slate gray, gray, or purplish-gray, and show thin–medium-bedded formation. The disappearance and emergence of shales (mudstones)

mark the base and top boundaries of the formation respectively. The thickness of the formation is 20.6 m.

Cephalopods are the most abundant fossils in this formation, followed by conodonts, brachiopods, and trilobites. Among them, the representative cephalopods are *Dideroceras wahlenbergi*, which are produced in the lower section of the formation and are closely related to the underlying Dawan Formation. The representative fossils of the upper section are *Ancistroceras* and *Paradnatoceras*. According to Ni Shizhao et al. (1983), the conodont fossils can be devided into the upper *Eoplacognathus fohaceus* Zone and the lower *Amorphognathus variabilis* Zone. The brachiopods produced in this formation include *Yangtzeella*, *Nereidella*, *Skenidioides*, etc. Trilobites include *Remopleurides*, *Nileus*, *Illaelus*, *Asaphus*, *Megalaspides*, *Birmanites*, *Lonchodomaas*, etc.

6) The Miaopo Formation ($O_{2-3}m$)

The Miaopo Formation is evolved from "the Miaopo shales" created by Zhang Wentang et al. (1957). The founding location of the formation was at Miaopo, Fenxiangchang, Yichang. It was originally called the Miaopo Shale Formation by Zhang Wentang (1962). Hubei Regional Surveying and Mapping Team (1970) called it the Miaopo Formation. Lu Yanhao et al. (1976), Hubei Three Gorges Strata Team (1978), and Zhang Jianhua et al. (1992) expanded the lower boundary of the Miaopo Formation to include all strata above the limestones at the top of the Guniutan Formation. However, most scholars at present follow the definition of Zhang Wentang et al. (1957) by using the appearance and disappearance of shales to mark the base and top boundaries of the Miaopo Formation.

The Miaopo Formation, in conformable contact with the two sets of carbonate strata in the respective Guniutan and Baota Formations, is composed mainly of yellowish-green and grayish-black calcareous mudstones, silty mudstones, and yellowish-green shales interbedded with thin-bedded bioclastic limestone lens. The formation contains abundant graptolite fossils as well as trilobite and cephalopod fossils. It has clear lithologic boundaries with the upper and lower strata, with the appearance and disappearance of mudstones (shales) marking the base and top boundaries respectively. The formation has a thickness of 3.1–6.6 m.

This formation is rich in graptolite, trilobite, ostracoda, brachiopod, cephalopod, and conodont fossils. Among them, the graptolite fossils can be divided into two zones: the upper *Nemagraptus gracills* Zone and the lower *Glyptograptus teretiusculus* Zone. The trilobites contained in this formation are mainly *Birmanites nileus*, *Telephina lonchodomas*, *Atractpyge*, *Tangyaia illaenus*, *Reedocalymane*, *Bumatus*, etc. The conodonts in the formation can be devided into three zones from bottom to top: ①the *Pygodus serra* Zone, ②the *P. anserinus* Zone, and ③the *Prioniodus alobatus* Zone. The cephalopods are mainly *Lituites*, *Cyclolituites*, etc. According to those fossils identified, this formation was formed during the late period of Middle Ordovician–the early period of Late Ordovician.

7) The Baota Formation (O_3b)

The name of the Baota Formation is evolved directly from "the Baota limestones" created by Li Siguang et al. (1924). This is the only stratigraphic unit in China that was named according to the morphological features of fossils. The founding place of this formation was at Leijiashan (once called Aijiashan), close to Longmaxi, 3 km east of Xintan in Zigui, Hubei. Yang Jingzhi and Mu Enzhi (1954) defined the Baota limestones as these limestones that contain *Sinoceras chinense*, which are about 13 m thick. Zhang Wentang et al. (1957) separated the 5 m-thick black shales that contain the *Glytoglutes teretiusculus* Zone at the top of Yang & Mu's Aijiashan formation and renamed them as "the Miaopo shales". Therefore, "the Baota limestones" is limited to the *Sinoceras*-bearing greenish-gray thin-bedded argillaceous limestones and dark purple cracked limestones that exist between the Miaopo shales and the Linxiang limestones. Later scholars have basically followed this definition.

The Baota Formation is composed mainly of shrinkage crack-bearing micrites associated with nodular micrites, which are gray, light purplish-red, or gray purplish-red and show medium-thick-bedded formation. With a thickness of 8.4–18.4 m, the formation is characterized by the production of cephalopods—*Sinoceras sinensis*.

This formation contains abundant cephalopod, conodont, trilobite, brachiopod, and ostracod fossils. Among them, the typical cephalopods contained in this formation are *Sinoceras chinense*, *Elongaticeras*, *Eosomichilinoceras*, and *Dongkaloceras*. The representative conodonts in the formation are *Hamarodus europaeus* and *Protopanderodus insculptus*. Trilobites are mainly found in the middle and upper sections of the formation, with *Paraphillipsinella globosa* typical in the middle section and *Nankinolithus* typical in the upper section. According to the characteristics of fossils, this formation should be formed during Late Ordovician.

8) The Wufeng Formation (O_3w)

The Wufeng Formation is evolved from "the Wufeng shales" created by Sun Yunzhu (1931). The founding place of the formation was at Yuyangguan, Wufeng, Yichang. The formation can be divided into the Graptolite Shale Member and the Guanyinqiao Member.

(1) The Graptolite Shale Member (O_3w^s). This member is equivalent to the original Wufeng shales named by Sun Yunzhu (1931) or the Wufeng Formation named by Zhang Wentang (1962). The member, with a thickness up to 5.44 m, consists mainly of organic content- and quartz silt-bearing sandy hydromica claystones assocated with blackish-gray, thin–very thin-bedded microcrystalline silicites, which are blackish-gray but show yellowish-green, light purple, or yellowish-brown after wheathering, and thin-very thin-bedded formation.

(2) The Guanyinqiao Member (Bed) (O_3w^g). The strata in this member were first discovered by Zhang Mingshao and Sheng Xinfu (1958) above the Wufeng shales, 2 km away from the south of the Guanyinqiao Bridge, Qijiang, Sichuan. Lu Yanhao (1959) called it "the Guanyinqiao Bed". Zhang Wentang (1964) changed it to "the Guanyinqiao Formation". Sheng

Pingfu (1971), Wang Ruzhi (1981), and Zeng Qingsheng (1983) renamed it as "the Guanyinqiao Member".

The lithology of the Guanyinqiao Member can be divided into three sections: At the lower section are blackish-gray, yellowish-brown, or light purplish-gray quartz-bearing silts and hydromica claystones. At the middle section are yellowish-gray, beige, or light purplish-gray quartz-bearing hydromica claystones (or rhyolitic tuffs). At the upper section are yellowish-gray of light gray hydromica claystones. The *Hirnantia* crustacean fauna in this member is characterized by the production of a large number of carpal faunas (represented by *Hernantia-Kinnella*) and trilobites faunas (represented by *Dalmanitina*). Altogether, there are nearly 35 species (across 28 genera) of brachiopods produced in this member including *Hirnantia*, *Dalmanella*, *Kinnella*, *Paromalomena*, *Eostropheodonta*, *Plectothyrella*, *Hindella*, etc. There are also productions of trilobites such as *Dalmanitina*, *Platycoryphe*, and *Leonaspis*.

The Wufeng Formation contains abundant graptolites, amounting to more than 200 species across 30 genera. In total, three graptolite zones and one crustacean fauna have been established. From top to bottom, they are:

(1) The *Normalograptus persculptus* graptolite zone, which has the first appearance of *Normalograptus persculptus* as the mark of its base boundary, and the first appearance of *Akidograptus ascensus* as the mark of its top boundary. The main paragenetic fossils in this member are *N. caudatus*, *N. madernii*, *N. wangjiawanensis*, *Glyptograptus laciniosus*, etc. In addition, all fossils in this member are double-row clinopyrites, with a majority being *Normalograptus*. Altogether, there are 29 species across 10 genera of *Normalograptus* produced in this zone. The top of this zone marks the top boundary of the Hirnantian.

(2) The *Hirnantia* shell fauna.

(3) The *Normalograptus extraordinarius* graptolite zone, which mainly consist of *N. ojsuensis*, *P. uniformis*, and *Neodiplograptus modestus*. In the deep-water facies of the zone such as Yichang, 32 species across 11 genera are found, among which up to 9 types of *Normalograptus* are identified, constituting the major genus of the zone. The base of this zone is taken as the base of the Hirnantian.

(4) The *Paraorthograptus pacificus* graptolite zone, which can be further divided into three subzones. One is the subzone of *Diceratograptus mirus*, whose base boundary is consistent with that of the *Paraorthograptus pacificus* graptolite zone, and the top boundary is marked with the appearance of *Tangyagraptus typicus*. The paragenetic molecules in this subzone are *Pararetiograptus parvus*, *Pseudoreteograptus nanus*, and *Paraplegmatograptus uniformis*, besides *P. brevispinus*, *Leptograptus planus*, *Dicellograptus tumidus*. The second subzone is the *Tangyagraptus typicus* subzone, which has the first appearance of *Tangyagraptus typicus* as the mark of its base boundary, and *Diceratograptus mirus* as the mark of its top boundary. The paragenetic fossils in this subzone are *T. remotus*, *T. flexilis*, *Dicellograptus*

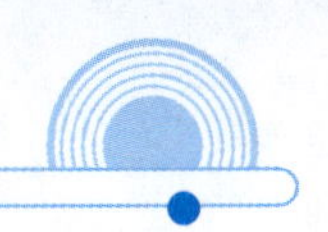

mirabilis, *Normalograptus angustus*, *N. normalis*, etc. And the last subzone is still unnamed, and thus called the unnamed subzone. This base boundary of the subzone is consistent with the base boundary of the *Paraorthograptus pacificus* graptolite zone, and its top boundary is marked with the first appearance of *Tangyagraptus typicus*. The paragenetic fossils in this subzone are *Pararetiograptus parvus*, *Pseudoreteograptus nanus*, and *Paraplegmatograptus uniformis*, besides *P. brevispinus*, *Leptograptus planus*, and *Dicellograptus tumidus*.

3. The Silurian System

The latest division plan divided the Silurian system into four sub-systems: the lower sub-system which is further divided into the lower Llandovery Series and the upper Wenlock Series, and the upper subsystem which is futher divided into the lower Ludlow Series and the upper Pridoli Series. The Silurian strata in the fieldwork area belong to the Llandovery Series of the lower subsystem.

1) The Longmaxi Formation (S_1l)

The Longmaxi Formation is evolved from "the Longma shales" created by Li Siguang and Zhao Yazeng (1924). The site is at Longmaxi, Xintan, Zigui. In this book, the Longmaxi Formation refers to a set of strata under the yellowish-green shales of the Xintan Formation and obove the Guanyinqiao Bed (Member), marked with the first appearance of *Akidograptus ascensus*. The formation is composed of black and grayish-green thin silty mudstones and quartz siltstones intercalated occasionally with thin fine quartz sandstones and yellowish-green silty mudstones and argillaceous siltstones, occasionally mixed with calcareous mudstone lens, and containing brachiopod and trilobite fossils. It is often weathered to light red–brownish-purple and purplish-gray in the region. The formation was named according to its founding place, a section at the Longmaxi, Xintan, Zigui. The Longmaxi Formation produces a large number of graptolite fossils, which can be divided into four graptolite zones from bottom to top:

(1) The *Akidograptus ascensus* graptolite zone. The zone takes the first appearance of *Akidograptus ascensus* as the marking of the base boundary, and the first appearance of *Parakidograptus acuminatus* as the marking of the top boundary. The symbiotic molecules mainly include *Glyphograptus* sp., *Neodiplograptus bicaudatus*, etc. In total, there are 22 species of graptolites across 6 genera in this zone.

(2) The *Parakidograptous acuuminatus* graptolite zone. The zone takes the first appearance of *Parackigrabus acuuminatus* as the marking of the base boundary and the first appearance of *Cystograptus vesiculosus* as the marking of the top boundary. The paragenetic molecules mainly include *Normalograptus premedius*, *N. rectangularis*, *Pseudorthograptus illustris*, etc. In total, there are 42 species across 12 genera in this zone. Among them, 6 genera and 24 species are new. The substantial presence of new species indicated a big change of graptolite

fauna occurred in this zone.

(3) The *Orthograptus vesiculosus* graptolite zone. The zone takes the appearance of *Ortoglobus vesiclosus* as the marking of its base boundary, and is also characterized by the substantial presence of *Ortoglobus vesiclosus*, associated with the appearance of *Dipiograptus modestus*, *D. longiformis*, *Climacograptus rectanguiaris*, and *C. normalis*.

(4) The *Coronograptus cyphus* graptolite zone. The zone is started by the appearance of *Coronograptus cyphus* or the production of *Pseudopemerograptus revolutus, Pernerograptus austerus, Monoclimacis lunata*, etc., and ended by the appearance of Demimurius triangulatus in the overlying zone (the Xintan Formation).

2) The Xintan Formation (S_1x)

The Xintan Formation evolved from "the Xintan shales" created by Blackwell (1907). Li Siguang (1924) called "the Xintan System" and further divided it into "the lower Longma shale" and "the upper Xintan shale". Xie Jiarong and Zhao Yazeng (1925) expanded the upper boundary of "the Longma shale" to a thickness of 400 m. Yu Jianzhang and Shu Wenbo (1929) cited "the Xintan System". Since 1959, in Hubei, the expanded Xintan System of the Longmaxi Formation (Group), the Luozaoping Formation (Group), and the Shamao Formation (Group) represent the Silurian System. In 1977, Hubei Bureau of Geology and Mineral Resources introduced the term "the Xintan Formation" in the lithostratigraphic cleanup, but revised its meaning to a set of yellowish-green shale, sandy shale, and thin siltstone intercalated with a small amount of thin-bedded fine sandstones between the Longmaxi Formation and the Luoreping Formation, with developed ripple marks and graptolite-producing stratigraphic sequence.

The Xintan Formation is rich in graptolites, with a majority in the lower part and a gradual decrease upward. It also has a small amount of trilobite and brachiopod fossils. According to the research by Wang Xiaofeng et al. (1987), three graptolite zones have been established from bottom to top: ①the *Demirastrites convolutus* Biozone; ②the *Monoraptus sedgwickii* Biozone; ③the *Coronograptus*? *arcuatus* Biozone. The third zone can extend upward to the lower part of the Luoreping Formation. Therefore, the age of this formation is generally Early Silurian.

3) The Luoreping Formation (S_1lr)

The Luoreping Formation is evolved from "the Luoreping System" created by Xie Jiarong and Zhao Yazeng (1925). The Luoreping Formation is located in Luoreping (also known as Dazhongba), Yichang. Yin Zanxun (1949) reclassified the section of Shamao Mountain, naming the 3–12 layers as the Luoreping Group. Mu Enzhi (1962) defined the Luoreping Group as the 7–12 layers of the section and attributed it to the Middle Silurian. It was renamed as the Luoreping Formation by Hubei Three Gorges Strata Team (1978), Ge Zhizhou et al. (1979), Mu Enzhi et al. (1982), and Wang Xiaofeng et al. (1987).

The Luoreping Formation refers to the stratigraphic sequence integrated between the

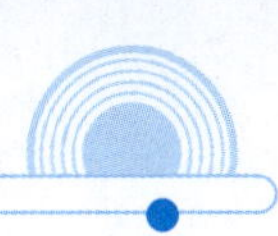

Xintan Formation and the Shamao Formation. The lower section is yellowish-green mudstone, shale intercalated with biological limestone, marl, or lens, producing fossils of brachiopods, graptolites, ect. The middle section is an interbedded layer of yellowish-gray mudstone, calcareous mudstone, and limestone or marl, producing crustaceous biota such as corallines and brachiopods. The upper section is yellowish-green mudstone and silty mudstone without limestone. The base layer starts with the appearance of limestone and the top layer ends with the base of sandstone. The thickness of the formation is 73.7–172 m.

The lower section of this formation contains abundant fossils of various categories, of which the brachiopods are represented by *Meifodia lissatrypaformts, Lisatrypa magna, Stricklandia transversa, Pentamerus robustus, P.* (*sulcupentamerus*) *hubeiensis, Apopentamerus hubeiensis, Katastrophomena depresa, Isorthi* sp. and so on; the corals are represented by *Palaeofauosites paulus, Favosites kogtdaensis, F. gothlandicus, Heliolites saiairicus, Onychophyilum pringlei, Halysttes* (*Acanthokalysites*) *pycnoblastoidu yabei, Pycnatis elegans*, ect.; the trilobites are represented by *Scotokarpes sinensis, Sckaryio hubeiensis*, etc; the graptolites are represented by *Monoclmacis arcuata, Giyntograptus sinuatus, Pseitdociimacograptus enskiensh* etc.; as well as conodonts, crinoid stems, polyzoa, cephalopods, bivalves, gastropods, etc. The fossils of the upper section are relatively dull and monotonous, among which the graptolites are *Climacograptus nebula, Pristiopograptus variabili sy, Oktavites planus*, ect.; the brachiopods are *Katastrophomena maxima, K. depresca, Lsorthis* sp. etc.; the trilobites include *Latiproetus latilimbatus, Luojaskanta xvangjiavanensis*, etc.; together with fish fossils (*Sinacanthus*), bivalves, gastropods, etc. This formation was formed during the middle–late periods of Early Silurian.

4) The Shamao Formation ($S_{1-2}sh$)

The Shamao Formation is evolved from "the Shamaoshan layer" created by Xie Jiarong and Zhao Yazeng (1925). The Shamao Formation is located in Luoreping, Shamao Mountain of Yichang. Yin Zanxun (1949) and Mu Enzhi (1962) defined the 13–20 layers of Maoshan section as "the Shamao Formation". Nangusuo (1974) named the lower and middle sections of the Shamao Formation as the Shiwuzi Formation and renamed the upper section as the Shamao Formation. Hubei Three Gorges Strata Team (1978) incorporated the Shiwuzi Formation into the upper part of the Luoreping Formation, belonging to Early Silurian, and the upper part is the Shamao Formation. Ge Zhizhou et al. (1979) also restored the Shamao Formation from the Shiwuzi Formation and divided it into the lower, the middle, and the upper parts. Wang Xiaofeng et al. (1987) followed the division opinion of Ge et al. (1979), renamed the Shamao Formation, and divided it into four segments. The first–third segments were classified as the Lower Silurian, and the fourth segment was classified as the Middle Silurian. This classification is followed in this book.

The Shamao Formation refers to the stratigraphic sequence which is integrated on the

yellowish-green silty mudstone of the Luoreping Formation and in parallel unconformity below the grayish-white thick-bedded quartzite-like sandstone of the Yuntaiguan Formation. The lower section is composed of yellowish-green shale, argillaceous siltstone, siltstone intercalated with sandstone or purplish-red fine sandstone. The upper section consists of grayish-green intercalated with purplish-red medium–thick-bedded fine-grained quartz sandstone mixed with medium–thin-bedded siltstone and sandy shale. It produces brachiopod, trilobite, bivalve, and other fossils.

This formation produces graptolites in the lower part of the stratotype section, with the main molecules being *Monograptus marri, M.* cf. *drepanoformis, Pristiograptus regularris, Pr. variabis,* etc., as well as trilobites, brachiopods, and conodonts, e.g., *Pterospathodus* cf. *celloni, Carniodus carnus, C. carnudus,* etc. In the middle, there are mainly brachiopods such as *Nalivkinia* cf. *elongata, Eospirifer* sp., and *Isorthis* sp., and trilobites such as *Coronacephalus* sp. and *Latiproetus* sp. The upper fossils are rare, with Brachiopoda *Strispirifer* sp. Judging from the above fossils, it belongs to the middle and upper parts of Llandovery Series in Early Silurian.

4. The Devonian System

1) The Yuntaiguan Formation ($D_{2-3}y$)

The Yuntaiguan Formation was founded by Yu Jianzhang and Shu Wenbo (1929), the origin name is "the Yuntaiguan quartzites", the site is in Yuntaiguan, South Dongqiao, Zhongxiang (under the jurisdiction of Dakou Farm in Zhongxiang City now). Yue Xixin (1948), and Yang Jingzhi and Mu Enzhi (1951, 1953) called it the Yuntaiguan quartzites, later Wang Yu and Yu Changmin (1962) renamed it as the Yuntaiguan Formation, which has been used until today.

The Yuntaiguan Formation is a set of grayish-white medium–thick-bedded or massive quartzite-like fine-grained quartz sandstone with a few grayish-green argillaceous sandstones. The formation sometimes mixed with purplish-red or fleshy-red thin-bedded siltstone or mudstone. At the bottom, there is basal conglomerate, pebbly sandstone or clay rock. Parallel unconformity occurs on different layers of the Silurian strata, in the Huangjiadeng Formation, or under the Dapu Formation, the Huanglong Formation, or the Liangshan Formation. The Yuntaiguan Formation generally contains few fossils. The thickness is 85.9 m.

2) The Huangjiadeng Formation (D_3h)

The Huangjiadeng Formation is evolved from "the Huangjiadeng layer" created by Yang Jingzhi and Mu Enzhi (1951). The place is in Huangjiadeng, the east wall of Ma'an Mountain, Changyang County. Later, Yang and Mu (1953) formally described the HuangJiadeng section through further studies. Wang Yu and Yu Changmin (1962) renamed it as the

Huangjiadeng Formation, and then it has been accepted used all the time.

The Huangjiadeng Formation is mainly composed of yellowish-green and grayish-green shale, sandy shale, and sandstone, sometimes intercalated with oolitic hematite deposits, containing fossils such as plants and brachiopods. It is in conformity with the underlying pure quartzite-like sandstone of the Yuntaiguan Formation and the marlstone and limestone overlying the bottom of the Xiejingsi Formation. The upper and lower boundaries are clearly distinguishable. Due to denudation, it can also be buried under the Dapu Formation, the Huanglong Formation, the Liangshan Formation, or the Qixia Formation, respectively, and its age belongs to Late Devonian. The Huangjiadeng Formation is the main ore-bearing stratum of Ningxiang-type iron deposits. In Songzi, Yidu, and Changyang areas, 1–3 layers can generally be seen, sometimes up to 4 layers, in a layered or lenticular shape. Among them, the top layer of this formation is relatively good, generally 1–3 m thick, with a maximum thickness of 11 m. The ore layer varies greatly along the lateral direction, and often transforms into iron sand shale. The above areas have a decreasing trend or pinching out in terms of layer number and thickness towards the west and east. The thickness is 12.8–15 m.

This formation contains relatively abundant animal and plant fossils. Among them, plants are *Leptoploeum rhombicum*, *Cyclostigma kiltorkense*, *Archaeopteris macilenta*, *A. fissilis*, *Rhacophyton ceratangium*, *Lepidodendropsis*? *arborecens*, etc.; brachiopods include *Cyrtospirifer anossafioids*, *C. pellizzariformis*, *C. sinensis*, *Lepotodema* cf. *naviformis*, *Tenticospirifer* sp., etc; fish fossils include *Changyanophyton hupeiense*. To sum up, the geological age of this formation is the early stage of Late Devonian, which belongs to marine-continental deposit. The lower part is dominated by continental facies, and the upper part is dominated by marine facies.

3) The Xiejingsi Formation (D_3C_1x)

The Xiejingsi Formation is evolved from "the Xiejingsi iron layer" created by Xie Jiarong and Liu Jichen (1929). The location is in Xiejingsi, Yidu (known as Zhicheng City now). After further investigation and research by Yang Jingzhi & Mu Enzhi (1951, 1953) and Jiang Tao (1965), a clear sequence and meaning were obtained. Hubei Bureau of Geology and Mineral Resources (1996) incorporated the narrowly defined the Xiejingsi Formation and the so-called "the Tizikou Formation" on the upper layer into the Xiejingsi Formation on the basis of basically consistent rock assemblages.

The Xiejingsi Formation refers to a set of stratigraphic sequences integrated between the Huangjiadeng Formation and the Dapu Formation. The upper part is called the sand shale section, mainly composed of grayish-green and grayish-black shale, carbonaceous shale, siltstone, and sandstone, occasionally containing oolitic chlorite, siderite, and coal lines, and containing brachiopods and plant fossils. The lower part is called the limestone section, mainly composed of gray and dark gray marlstone, limestone, or dolomite, sometimes mixed with shale and oolitic hematite or oolitic chlorite siderite, containing brachiopod fossils. The area is caused by denudation, and the overlying strata vary from place to place. The thickness of the formation is 11.66 m.

The lower limestone section of the Xiejingsi Formation is rich in brachiopods, as well as fossils of bryozoans, corals, ostracods, etc. Among them, there are brachiopods, such as *Yunnanella abrupta*, *Y. simplex*, *Y. zuangi*, *Yunnanellina hanburyi*, *Y. hunanensis*, *Y. simplex*, *Cyrtospirifer chaoi*, *C. davidsoni*, *C. pellizzarformis*, *Tenticospirifer hayasaki*, *T. Tenticuium*, *Producttlla subbacidiatus*, *Athyis gurdoni*, and *Hunanospirifer ninghsiangensi*; corals, such as *Billingsastraea* sp. and *Pseudozaphrontis curulena*; bryozoa, such as *Rhombopora* and *Leptotrypa* sp.; and ostracodes, ect. Therefore, this part of the stratohorizon undoubtedly belongs to Late Devonian. The upper sand shale section of the Xiejingsi Formation contains plant fossils and spores. Among them, plant fossils mainly include *Hamatophyton verticilliatum*, *Leptoptoptokloeum rhombicum*, *Cyclostigma kiltorkense*, *Lepidodendropsis hirmeri*, *L. Theodori*, *Archaeosigillaria vanuxemi*, *Preleptdodendron yiduense*, *Pseudobombia ursine*, *Barinopkyton citrulliforme*, *Drepanophycus spinaeformis*, *Eolepidodeodoron wusihense*, *Subiepidodendron mirabie*, etc. From the characteristics of the plant fossils mentioned above, this part of stratohorizon still belongs to Late Devonian. However, the upper part of this section contains relatively abundant brachiopods, ostracods, conodonts, and a small amount of corals, among which the main brachiopods are *Schucherteua gelaohoensis*, *S. guizhouensis*, *Leptagonia analoga*, *Spirifer attenuatus*, *Cruithyris urei*, *Ptychomanetochia kinlingensis*, *Camarotoeckia kinlingensis*, etc.; conodonts are *Leiiognathus levis*, *Polygnathus inomatus*, *Pseudopolygrtatkus originalis*, ect.; and a small amount of corals like *Syringpora ramulom*, ect. According to this, the stratum in this section should belong to Early Carboniferous, indicating that the upper sandstone section of the Xiejingsi Formation is an age-acrossed strata.

5. The Carboniferous System

1) The Dapu Formation (C_2d)

The Dapu Formation evolved from "the Dapu dolomite layer" created by Zhang Wenyou (1944). The site is in Liucheng, Guangxi (formerly known as Dapu Town in Liucheng). In Hubei, this stratum has always been included in the lower part of the Huanglong limestones (陈旭，1935；岳希新，1948；杨敬之，1954), or the lower part of the Huanglong Group (杨敬之等，1952；湖北区域地质测量大队，1965–1975), or the lower part of the Huanglong Formation (湖北区域地质测量大队，1984；顾威国，1982). Feng Shaonan et al. (1984) classified this set as the lower dolomite section of the Huanglong Formation and assigned it to the Dapu Formation, which was adopted by Hubei Bureau of Geology and Mineral Resources.

The Dapu Formation refers to a set of grayish-white–grayish-black thick-bedded

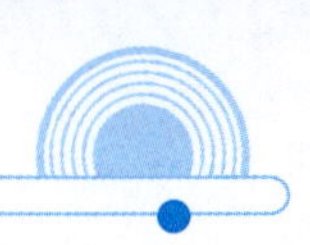

dolomites that are parallel unconformable above the Xiejingsi Formation and below the limestone of the Huanglong Formation, and are bounded by the disappearance of thick-bedded dolomites or the occurrence of limestone above. The thickness is 5.1 m.

The fossils in this formation are relatively rare, and fossils like non-fusulinid foraminifera, fusulinids, corals, etc. can only be obtained in some interbeds composed of dolomitic limestones or limy dolomites which contain higher gray matter. Among them, the non-fusulinid foraminifera are *Glomospira vulgaris*, *Tolypammina fortis*, etc.; the fusulinella includes *Profusulinella* cf. *marblensis*, *Eofusulina* cf. *trianguliformis*, *Pseudostaffella composita keltmica*, etc.; the corals includes *Lophophyllidium* sp., etc. This indicates that the formation belongs to the early Late Carboniferous.

2) The Huanglong Formation (C_2h)

The Huanglong Formation evolved from "the Huanglong limestones" created by Li Siguang and Zhu Sen (1930). The site is located in the west part of Huanglong Mountain, Longtan, Nanjing, Jiangsu. The Huanglong limestone was first divided by Chen Xu (1935) from the Yangxin limestone named by Xie Jiarong (1924) in the southeast of Hubei. Yang Jingzhi and others (1962) renamed it as "the Huanglong group" which has been used until the early 1980s. Later, Gu Weiguo (1982) and Feng Shaonan (1984) designated the Dapu Formation as the lower section of the Huanglong Formation in the Huangshi area of eastern Hubei and the lower section of the former Huanglong Formation in the Three Gorges area of western Hubei respectively. Then the Huanglong Formation only refers to the upper pure limestone, which is the meaning of the Huanglong Formation adopted by Hubei Bureau of Geology and Mineral Resources (1996).

The Huanglong Formation is a set of gray, light gray, and fleshy-red thick-bedded microcrystalline limestone and bioclastic limestone, with a coarse crystalline limestone at the bottom, containing calcareous dolomite breccia and conglomerates, and rich in corals, brachiopods, and other fossils. The upper sandstone is in parallel unconformable contact with sandstone of the Permian Liangshan Formation, and the lower sandstone is in conformable contact with fine crystalline dolostone of the Dapu Formation.

This formation contains abundant fossils of non-fusulinid foraminifera, fusulinids, brachiopods, corals, etc. Among them, the non-fusulinid foraminifera from bottom to top is represented by *Tolypamina fortis*-*T. hubeiensis* Assemblage Zone and the *Bradyina minima*-*Plectogyra minuta* Assemblage Zone. The fusulinella species are extremely abundant and are characterized by *Fusulinella*, *Fusulina*, *Beedeina*, *Fusiella*, *Pseudostaffella*, and other genera. The brachiopods are *Ella simensis*, *Athyris planosulcata* var. *uralica*, and *Neochonetes carbonifera*. The corals are mainly *Chautetes*, *Caninia*, etc. From the perspective of fusulinids and non-fusulinid foraminifera, the age of this formation is the early period of Late Carboniferous.

6. The Permian System

1) The Liangshan Formation (P_2l)

The Liangshan Formation evolved from "the Liangshan stratum", named by Zhao Yazeng and Huang Jiqing (1931). The place is in Zhongliang Temple, Liangshan, Nongfeng, Nanzheng, Shanxi. The strata layer is called "the coal series at the base of Yangxin limestone" by Xie Jiarong & Liu Jichen (1927) and Yu Jianzhang & Shu Wenbo (1929). Li Jie et al. (1937) created the name "Ma'anshan coal-bearing series" in Western Hubei. Gao Zhenxi and Chu Xuchun (1940) gave the name "the Matupo coal system" in Podong. Yang Jingzhi and Mu Enzhi (1954) "the Ma'an coal system". Beijing Collgeg of Geology (1960) called it "the Liangshan Formation" and has been used since then.

The Liangshan Formation refers to the rocks which are in unconformity parallel contact with the Huanglong Formation in upper sections and integrate contact with the overlying Yangxin Formation. The lower section is composed of grayish-white medium–thick-bedded quartzite-like fine sandstone, siltstone, mudstone, and coal seam. The upper section is black thin mudstone intercalated with limestone lens. The thickness is 3.8–4.2 m.

The Liangshan Formation contains abundant fossils, including plant fossils, such as *Sigillaria acutaguia*, *Lepidodendron oculusfelis*, *Stigmaria ficoides*, *Pecopteris* sp., and *Sphenopteris* sp.; brachiopods, such as *Orthotiichia magnifica*, *Ogbinia hexaspinom*, *Tyloplecta richthofeni*, *Neochonetes nantanensis*, *Plicatifera minor*, and *Ambococelta* sp.; and ostracodas, like *Hollinella liangshanensis*, *Rimndydla hubeiensis*, etc. According to the fossils and stratigraphic relationships mentioned above, its geological age is the early period of Late Permian.

2) The Qixia Formation (P_2q)

The Qixia Formation is evolved from "the Qixia limestones" created by Richthofen (1912). The site is located in Qixia Mountain, the eastern suburb of Nanjing. It refers to a set of carbonate rocks between the Liangshan Formation and the Gufeng Formation (or the Maokou Formation). It was in parallel unconformity contact with the underlying Huanglong Formation (or in integration contact with the underlying Liangshan Formation) and in integration contact with the shale containing manganese or phosphorous nodules of the overlying Gufeng Formation (or in integration contact with the overlying Maokou Formation). The lithology of the Qixia Formation in this intership area is relatively single, mainly consisting of a series of dark gray and grayish-black thick-bedded chert-bearing bioclastic micrite. The dark gray thick knobby bioclastic micrite only develops on the top and base, and the limestone in base layer is sandwiched with calcareous carbonaceous shale. The thickness of the formation is 88.9 m.

This formation is rich in biological fossils, especially in areas like Zigui–Xingshan and the southeastern Hubei. Fusulinids include *Nankinella orbicularia*, *N. globularis*, *Sphaerulina hunanica*, and *Pisolina ercessa* are produced; corals include *Wentzellophyllum volzi*, *Cystomichelinia* sp., *Hayasakaia elegantula*, *Polytecalis yangtzeensis*, *P. chinensis*, etc. Brachiopodas include *Orthotichia chekiangensis*, *Tyloplecta richthofeni*, etc. Others are bryozoa, ostracods, conodonts, etc. According to this, its geological age of this formation is the early and middle periods of Middle Permian.

3) The Maokou Formation (P_2m)

The Maokou Formation is evolved from "the Maokou limestones" created by Lesen Xun (1929). The site is located in the bank area of Maokou River in Langdai, Guizhou. After Xie Jiarong (1924) defined the name "the Yangxin limestone" of the Lower Permian in the southeastern Hubei, there were many people doing researches, such as Yue Xixin (1948), Yang Jingzhi and Mu Enzhi (1954), and Zhou Shengsheng (1956). Later, Sheng Jinzhang (1962) proposed a plan to divide the Lower Permian in the southern China into "the Qixia Formation" and "the Maokou Formation" from bottom to top, which has been widely used by Chinese geologists.

The Maokou Formation refers to the rocks which are in integrated contact with the dark gray chert-bearing limestones of the Qixia Formation and in parallel unconformity contact with the argillite at the bottom of the Longtan Formation (or the blackish-gray thin-bedded siliceous rock intercalated with siliceous mudstone of the Gufeng Formation). The lithology of the Maokou Formation is mainly a set of gray and light gray thick–massive bioclastic microcrystal limestone containing chert nodules, algal micrite, and bioclastic arenaceous sparite, with 2–3 layers of fine crystalline dolostone sandwiched in the middle. In the middle and upper limestone, there are often dense chert nodules or bands. From its lithofacies and biological studies, this area has features of open platform facies.

The Maokou Formation is rich in biological fossils, with the fusulinid belt including (established from bottom to top): the *Verbeekina grabaui* belt, the *Chusenella conicocylindrica* belt, the *Neoschwagerina haydeni* belt, and the *Yabeina* belt; and coral species including the *Ipciphyllum subtimoricum-I. eligantum* Abundance Zone, the *Tachylasma elongatum-Paracaninita liangshunsis* Abundance Zone, etc. Other biological categories are also abundant, such as bryozoa, brachiopod, and non-fusulinid foraminifera. According to this, the geological age is the late Middle Permian.

4) The Wuchiaping Formation (P_3w)

The Wuchiaping Formation is evolved from "the Wuchiaping limestones" created by Lu Yanhao (1956). The site is in Wujiaping, Northwest Nanzheng, Shanxi. After Shengjinzhang (1962) revised its definition, it refers to a stratigraphic unit which is below the Changshing

Formation and above the Maokou Formation. Since then, this definition has been established.

The Wuchiaping Formation refers to the stratigraphic sequence integrated between the Maokou Formation limestone and the Daye Formation marlstone, and it consists of gray medium–thick-bedded and massive chert-bearing marl and bioclastic limestone. A layer of oolitic iron-aluminum argillaceous rock (the Wangpo Member) is stably developed at the bottom, and the bottom of this layer is taken as the base boundary. The top boundary of this formation is the end of chert limestone or the occurrence of laminated limestone and thin-bedded argillaceous limestone. The thickness of the formation is 84–103 m.

This formation contains abundant marine benthic fossils, among which fusulinids, corals, and brachiopods have been studied systematically, and biostratigraphic units have been established respectively from bottom to top: the fusulinids, such as the *Codonofusiella* Abundance Zone and the *Palaeofusulina sinensis* Zone; corals, such as the *Plerophyllum guangxiense-P. Sintanense* Assemblage Zone and the *Waagenophyllum lui-Lophocarino phyllum* Assemblage Zone; brachiopods, such as the *Tshernyschewia sinensis-Loipinga ruber* Assemblage Zone and the *Squamularia grandis* Assemblage Zone. According to the biological fossils' characteristics listed above, this formation was formed during Late Permian.

2.1.2.3 *Mesozoic*

1. The Triassic System

1) The Daye Formation (T_1d)

The Daye Formation is evolved from "the Daye limestones" created by Xie Jiarong (1924). The site is in Tieshan Iron Mine, North Daye, Hubei. In the following year, Xie Jiarong and Zhao Yazeng quoted the name to the west of Hubei Province and classified the upper thin-bedded limestone of Wushan in Blackweider (1907) as the Daye limestones. Zhao Jinke et al. (1962) classified the dolomite-dominated strata in the upper section of Daye limestones in the western Hubei as "the Jialingjiang Formation" and attributed it to Middle Triassic. The limestone in the lower section, dominated by thin layers, is called "the Daye Formation" and attributed to the Lower Triassic or the Daye System. Since then, it has been widely used in South China and was called "the Daye Group" before 1970s and "the Daye Formation" after 1970s.

The Daye Formation is mainly composed of gray and light gray thin-bedded limestone. The middle and upper sections are sandwiched between medium–thick-bedded limestone, sometimes with oolitic limestone, dolomitic limestone, or dolomitization limestone, and the lower section contains argillaceous limestone or yellowish-green shale. The bottom boundary is integrated with the shale and the Wuchiaping Formation in the grayish-black

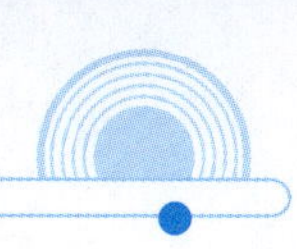

thick bedded and massive micritic–microcrystalline bioclastic limestones (containing medium chert nodules), and the top boundary is in contact with the overlying dolomites of the Jialingjiang Formation. The thickness of the formation is 1000 m.

This formation is mainly composed of ammonites, bivalves, and conodonts, and its lower section is rich in ammonites, mainly represented by *Ophiceras, and Lytophiceras*. The bivalves are *Claraia wangi, C. griesbachi*, etc. The conodonts are *Anchignathodus typicalis, Neogondolella carinata*, etc. The upper section is mainly composed of bivalves like *Eumorphotis multiformis, Bakevellia mediocatics minor, Leptochondria virgalensis*, etc., and the conodonts like *Neospathodus hubeiensis, Neohindeodella triassica*, etc. According to the above fossils, this formation was formed during Induan, Early Triassic.

2) The Jialingjiang Formation (T_1j)

The Jialingjiang Formation was originally called "the Zhaohua limestones" by Zhao Yazeng and Huang Jiqing in 1931 (赵亚曾, 1929), later renamed "the Jialingjiang limestones", and the site is in the coastal zone of Jialingjiang River, about 15 km from Guangyuan in the north. Zhao Jinke et al. (1962) cited it to the western Hubei according to the investigations given by Luo Zhili et al. (1957). Hubei Institute of Geological Science (1962) and Hubei Regional Surveying and Mapping Team (1966) cited it in the southeastern Hubei respectively, and called the Jialingjiang Formation (Group), which has been widely accepted since then.

The Jialingjiang Formation is mainly composed of gray medium–thick-bedded dolomite and dolomitic limestone with microcrystalline limestone and "evaporite solution breccia". It contains marine bivalves and foraminiferal fossils, and cephalopods are rare. It is in integrated contact with the gray thin limestone of the underlying Daye Formation and mottled mudstone and dolomite at the bottom of the overlying Batong Formation. The thickness of the formation is 728 m.

Since the strata of this formation are mainly characterized by dolomite, large fossils are generally rare. Fossils are mainly represented by bivalves, and a few ammonites and brachiopods. The microfossils are conodonts and foraminiferas. The lower section of this formation is composed of bivalves (like *Eumorphotis inaequicostata, Bakevellia exporrecta*, etc.), ammonites (like *Paragoceras sinense*), conodonts (like the *Pachycladina-Parachirognathus ethingtoni* Zone), and foraminiferas (characterized by *Aulotortus ckialingckiangensis*). The middle and upper sections are characterized by bivalves (like *Leptochondria minima*, *Chlamys weiyuanensis*, etc.), conodonts (like the *Neospathodua triangularis–N. homeri* Zone), and foraminiferas (like *Glomospira sinensis, Meandrospira insolita*). The top generally contains fossils represented by bivalves, e. g., the *Eumorphotis (Asoella) illyrica* Assemblage Zone. According to the above fossils, this formation from bottom to top formed during the late period of Early Triassic.

3) The Badong Formation (T_2b)

The Badong Formationis evolved from "the Badong layer" created by Richthofen (1912). The site is in the shores along the Yangtze River in Badong. The name was changed into "the Badong System" by Xie Jiarong and Zhao Yazeng (1925) and has been widely used by later generations. Zhao Jinke (1962) renamed it as the Badong Formation.

The lithology of the Badong Formation can be divided into three sections. The upper and lower sections are purplish-red siltstone and mudstone, sandwiched with grayish-green shale and occasionally containing malachite film. The middle section is limestone and marlstone. Grayish-green shale is commonly seen at the bottom and is in integrated contact with the underlying Jialingjiang Formation and the Jiuligang Formation of the overlying Xiangxi Group. The thickness of the formation is 75–91 m.

The formation mainly produces bivalves, as well as ammonites and plant fossils, which are mostly concentrated in the middle and lower section of the Badong Formation, followed by the bottom and rare in the upper section. The bivalves mainly include *Eumorphotis (Asoella) subillyrica*, *E. (A.) illyrica*, *Myophoria (Costatoria) goldfussi*, *M. (C.) submulthtriata*, and *M. (C.) goldfussi mansuyi*. The ammonites include *Progonoceratites* sp. The plant fossils include *Annalepis zeilleri*, etc. This indicates that the formation was formed during Middle Triassic.

2. The Xiangxi Group (T_3J_1X)

The Xiangxi Group is evolved from "the Xiangxi coal-bearing sandstone series" created by Noda Seijiro (1917) and the site is in Xiangxi, Zigui. Li Siguang et al. (1924) called it the Xiangxi System. Xie Jiarong and Zhao Yazeng (1925) called it the Xiangxi coal series; Si Xingjian and others (1962) renamed it as the Xiangxi Group. The lower coal formation of the Xiangxi Group was designated as the Rhaetian Stage of the Upper Triassic by Beijing Collgeg of Geology (1960); the middle and upper coal formations belong to the Lias Stage of the Lower Jurassic. Hubei Regional Surveying and Mapping Team (1973) and *Regional Geological Records of Hubei Province* (1990) have newly named the lower, the middle, and the upper coal formations of the Xiangxi Group in the western Hubei as the Jiuligang Formation, the Wanglongtan Formation, and the Tongzhuyuan Formation respectively. The first two formations were formed during Late Triassic and the latter was formed during Early Jurassic. Chen Chuzhen et al. (1979) created a new name for the lower coal formation of the Xiangxi Group in the Zigui Basin as "the Jiuligang Formation", dating from Late Triassic. The middle and upper coal formations are referred to as the Tongzhuyuan Formation (in a narrow sense), dating from Early–Middle Jurassic. Hubei Bureau of Geology and Mineral Resources (1996) called it the Xiangxi Group. In the Zigui basin, this group includes the Jiuligang Formation

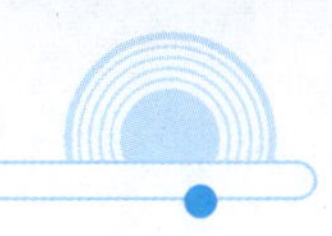

and the Tongzhuyuan Formation from bottom to top. In the Jingdang basin and the southeastern Hubei, the group includes the Jiuligang Formation, the Wanglongtan Formation, and the Tongzhuyuan Formation from bottom to top.

1)The Jiuligang Formation (T_3j)

The Jiuligang Formation is mainly composed of yellowish-gray and dark gray siltstones, sandy shales, and mudstones, mixed with feldspathic quartz sandstones and carbonaceous shales, containing 3–7 layers of coal seam or coal streak, with a total thickness of 41–142 m. This formation is continuously deposited with the overlying thick-bedded quartz sandstone of the Tongzhuyuan Formation and the underlying Badong Formation. However, in areas of Xiangxi River (Zigui) and Gengjiahe River (Xingshan), this formation is in direct contact with the lower section of the Badong Formation.

The flora of the Jiuligang Formation is dominated by cycads, and ferns are also well developed, especially for Pteridaceae. This flora has the dual characteristics of the northern flora and the southern flora. The main combinatorial molecules are *Lepidopteris-Pterophyllum bavieri and Drepanozamites-Cycadocarpidium*. The geological age is Late Triassic.

2) The Tongzhuyuan Formation (J_1t)

The Tongzhuyuan Formation is mainly composed of yellow, yellowish-green, and grayish-yellow sandy shales, siltstones, and feldspathic quartz sandstones, with carbonaceous shales and thin seam or coal streak, and a conglomerate layer at the bottom. It contains plant and bivalve fossils. This formation is in integrated contact with the underlying Jiuligang Formation and the overlying Qianfoya Formation. The thickness of the formation is 280 m.

The paleontology in this formation is characterized by the flora of *Coniopteris-Ptilophyllum contiguum-Sphenobaiera huangi* and the fauna of *Pseudocardonia-Qiyangia cuneata*. Therefore, it was formed during Early Jurassic.

3. The Jurassic System

1) The Qianfoya Formation (J_2q)

The Qianfoya Formation was named by Zhao Yazeng and Huang Jiqing in 1931 at Qianfoya, the north of Guangyuan and the east of Jialingjiang River. Formerly known as "the Tsienfuyen Formation". Chen Chuzhen et al. (1979) cited it in the Zigui basin, which is equivalent to cites given below: Xie Jiarong et al. (1925) first called it as the lower section of the Guizhou Formation, Beijing College of Geology (1960) called it the Ziliujing Formation, Hubei Regional Surveying and Mapping Team (1984) created the name of "the Niejiashan Formation", Zhang Zhenlai et al. (1987) created "the Qianfoya Formation" and "the Chenjiawan Formation". Hubei Bureau of Geology and Mineral Resources (1996) adopted the

Qianfoya Formation to refer to a set of stratigraphic sequences located between the Tongzhuyuan Formation and the Shaximiao Formation.

The bottom of the Qianfoya Formation is a layer of gravel-bearing quartz sandstone, sometimes the gravel is enriched into a thin bed and is marked as the bottom boundary, which is in integrated contact with greenish-yellow and gray calcareous mudstones of the Tongzhuyuan Formation of the underlying Xiangxi Group. The lower section is purplish-red and greenish-yellow mudstones, siltstones, and fine-grained quartz sandstones interbedded with shell limestones, rich in bivalve and sporopollen fossils. The upper section is mainly composed of purplish-red with yellowish-gray mudstones, sandy shales, siltstone, and feldspathic quartz sandstones. It is in integrated contact with the yellowish-gray massive lithic feldspathic sandstones at the bottom of the Shachimiao Formation. The thickness of the formation is 390 m.

This formation contains fossils of bivalves, plants, and sporopollen. It mainly producing bivalves, including *Pseudocardinia kweichuensis*, *P. longa*, *Lamprotula* (*Eolamprotula*) *solita*, *L.* (*E.*) *cremeri*, *Psilunio crvalis*, etc. Accordingly, this formation was formed during the early period of Middle Jurassic.

2) The Shaximiao Formation (J_2s)

The Shaximiao Formation is an evolution of "the Shaximiao layer" created by Yang Boquan and Sun Wanquan (1946) from the original "Chongqing Series" (Ha Anmu, 1931); the location is in Shaximiao of Hechuan, Sichuan. Xie Jiarong et al. (1925) once called it as the central section of the Guizhou System in the Zigui basin. This formation was first cited by Beijing College of Geology (1960) as the Shaximiao Formation of the Guizhou Group. Hubei Bureau of Geology and Mineral Resources (1996) called it the Shaximiao Formation.

The lithology of the Shaximiao Formation is yellowish-gray and purplish-gray feldspathic quartz sandstone, and purplish-red and purplish-gray mud (shale) rocks with unequal thickness and interbedded sequences, containing fossils of bivalves, ostracods, conchostraca, plants, and vertebrates. It is in integrated contact with the underlying Qianfoya Formation and the brick-red debris feldspathic sandstone at the bottom of the overlying Suining Formation, and can also overlap different layers of the Ziliujing Formation in parallel unconformity. The top boundary of "the Conchostracan shales" can be divided into two sections. The thickness of the formation is 1986 m.

The fossils in Shaximiao Formation are rare. *Chungkingithys xilingensis* was found at the bottom of the formation in Guojiaba, Zigui. The lower and upper sections contain the ostracodas, like *Darwinula* aff. *Sarytirmenensis*, *Clinocypris xilingensis*, and sporopollen assemblage, like the *Cyathidis-Classopollis-Neoristica* Assemblage Zone, etc. This formation was formed during the late period of Middle Jurassic.

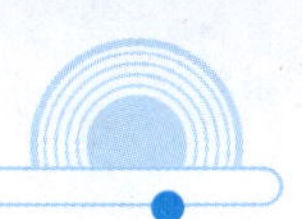

2.2 Sedimentary Rocks and Sedimentation

Sedimentary rocks in Zigui fieldwork area are widely distributed and are the main rock types of Neoproterozoic–Cenozoic strata in this area. The exposed sedimentary rocks are mainly terrestrial clastic sedimentary rocks from allogenic rocks and carbonate rocks in authigenic sedimentary rocks. The Neoproterozoic Nanhuaian series is mainly terrigenous clastic rocks, which are fluvial and glacial deposits deposited on the weathered and denuded surface formed by the Jinning Movement. Sinian (Neoproterozoic)–Ordovician (the Lower Paleozoic) is mainly composed of authigenic carbonate sedimentary rocks. Only in the Cambrian Shuijingtuo Formation–the Shipai Formation, there are more terrigenous clastic rocks. It is generally the sedimentary facies of the basin marginal-constrained sea platform. Sinian–Cambrian authigenic carbonate rocks are dominated by dolomite, and Ordovician is dominated by limestone. Silurian (the Lower Paleozoic)–Devonian (the Upper Paleozoic) is dominated by terrigenous clastic rocks, affected by the Caledonian Movement, and lacks the Middle and Upper Silurian and the Lower Devonian. Carboniferous (the Upper Paleozoic)–the Lower Triassic (Mesozoic) are dominated by authigenic carbonate deposits, mainly carbonate platform facies. Middle Triassic (Mesozoic)–Quaternary (Cenozoic) is dominated by terrigenous clastic rock deposits, mostly fluvial facies, lacustrine facies, and piedmont-diluvial facies.

The following is a brief summary of terrigenous clastic rocks and authigenic carbonate rocks commonly found in the fieldwork area.

2.2.1 Terrigenous clastic rocks

2.2.1.1 *Coarse clastic rocks*

The coarse clastic rocks exposed in the fieldwork area are mainly conglomerates.

At the bottom of the Nanhuaian Liantuo Formation, there is a thin layer of dark purplish-red pebblestone. The gravel content is about 30%–40%, the matrix is mainly fine sand, and shows matrix-support structure. The gravel composition is mainly granite and quartzite, the grain size is about 0. 5–2 cm, the sorting is medium–poor, the roundness is round–subround. From the bottom to the top of the profile, the size of the gravel becomes finer and the content decreases, becoming a river-channel stagnant sediment.

The conglomerates of the Nanhuaian Nantuo Formation are mainly conglomerates, formed by the deposition of glaciers and meltwater. The typical characteristics are: gravels are mixed in size, the larger ones can reach more than 50 cm, and the smaller ones are only a few centimeters; the shapes are diverse, with angular–subangular gravel and round–subround gravel; complex composition, igneous rock, sedimentary rock, and metamorphic rock can be seen.

At the bottom of the Jurassic Tongzhuyuan Formation is dark gray pebblestone. The gravel content can be as high as 70%, and the matrices are mainly silt–fine sands with grain-support structure. The gravels are mainly composed of siliceous rock and quartzite, with a grain size of about 3 cm, excellent in sorting, and the roundness is good.

At the bottom of the Cretaceous Shimen Formation is light gray–purplish-red boulderstone–cobbleston. The gravel content is up to 80%, the matrices are mainly medium–fine sands, with grain-support structure and calcium cementation. The gravel composition is dominated by limestone and dolomite, mixed in size, ranging from several centimeters to 20 cm, poor in sorting, and the roundness is subround–subangular (Figure 2-1).

pebblestone at the bottom of the Nanhua Liantuo Formation (Nh_1l) (Sixi)

conglomerate glacial of the Nanhua Nantuo Formation (Nh_2n) (Huajipo)

pebblestone at the bottom of the Jurassic Tongzhuyuan Formation (J_1t) (Guojiaba)

cobblestone at the bottom of the Cretaceous Shimen Formation (K_1s) (Zhouping)

Figure 2-1 Types of coarse clastic rocks in the fieldwork area

2.2.1.2 *Sandstones*

Sandstone in the fieldwork area is mainly found in the Nanhuaian Liantuo Formation, the Silurian Shamao Formation, the Devonian Yuntaiguan Formation, the Devonian Huangjiadeng Formation, the Triassic Jiuligang Formation, the Jurassic Tongzhuyuan Formation, and the Jurassic Qianfoya Formation (Figure 2-2).

purplish-red medium-grained quartz sandstone in the Second Member of the Nanhuaian Liantuo Formation (Nh_1l) (Huajipo; possibly delta deposition,visible large-angled cross-bedding)

fine-grained quartz sandstone of the Silurian Shamao Formation ($S_{1-2}sh$) (Zhouping)

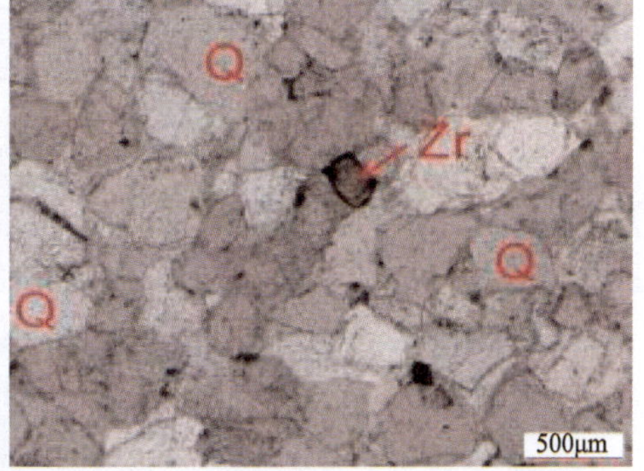

fine-grained quartz sandstone of the Devonian Yuntaiguan Formation ($D_{2-3}y$) (Zhouping)

Figure 2-2 Types of sandstone in fieldwork area

The sandstone developed in the Nanhuaian Liantuo Formation is purplish-red and grayish-green coarse–medium feldspathic quartz sandstone and feldspar sandstone in the middle–lower parts, and purplish-red and grayish-white crystalline debris or lithic tuffaceous sandstone and lithic sandstone in the upper part. This set of sandstones is mainly fluvial deposits with abundant horizontal bedding and cross bedding.

The Silurian sandstones in the Shamao Formation mainly appear in the upper part of the Shamao Formation. They are mainly grayish-green and grayish-white fine-grained quartz sandstone with cross-bedding and corrugated structures. In the lower and middle sections, the sandstones mainly appear in fine clastics as interlayers.

The Devonian Yuntaiguan Formation is mainly composed of grayish-white and flesh-red fine-grained quartz sandstone and feldspathic quartz sandstone. The quartz sandstone has a high degree of maturity in composition and structure. Cross-bedding can be seen in the field. Under the microscope, overgrowth of quartz can be seen, mainly for shore sediments.

The sandstone in the Devonian Huangjiadeng Formation is often interbedded with fine clastic rocks, and the main type is light gray fine-grained quartz sandstone.

The Triassic sandstone in the Jiuligang Formation is mainly grayish-yellow and grayish-green feldspathic quartz sandstone. The sandstone of the Jurassic Tongzhuyuan Formation is dark gray medium-grained quartz sandstone at the bottom, and in the middle and upper parts, grayish-yellow fine sandstone are interbedded with fine clastic rocks. The sandstone in the Jurassic Qianfoya Formation is grayish-yellow and fine-grained quartz sandstone and fine clastic rock symbiosis.

2.2.1.3 *Fine clastic rocks*

Fine clastic rocks mainly include siltstones and argillaceous rocks, which are widely develop in the fieldwork area (Figure 2-3). The Nanhuaian fine clastic rocks mainly deposit in the Nantuo Formation. The lithology is grayish-green and purplish-red glacial till-bearing gravel siltstone and glacial till-bearing silty mudstone, often constituting a basic sedimentary sequence with glacial conglomerate and glacial till-bearing siltstone.

Sinian fine clastic rocks are mainly found in the second and fourth members of the Doushantuo Formation. The fine clastic rocks in the second member of the Doushantuo Formation are mainly black and dark brown carbonaceous mudstone or shale, interbedded with argillaceous and carbonaceous dolomite, forming the basic sedimentary sequence of the second member of the Doushantuo Formation. The fine clastic rocks in the fourth member of the Doushanduo Formation are black carbonaceous shale, siliceous shale, and silty shale.

Cambrian fine clastic rocks are mainly found in the Shuijingtuo Formation and the Shipai Formation. The middle and lower rock sections of the Shuijingtuo Formation are dominated by fine clastic rocks, mainly blackish-gray and grayish-yellow carbonaceous shale and silty shale. The upper calcareous shale is mainly interbedded in the limestone rock. The fine clastic rocks of the Shipai Formation are appearing in the lower and upper layers. The lower layers are yellowish-green silty mudstone and siltstone, and the upper layers are purplish-gray and grayish-green silty shale and silty mudstone.

The most characteristic fine clastic rocks of the Ordovician appear in the Wufeng Formation, mainly composed of grayish-black and grayish-yellow siliceous mudstone, rich in graptolite fossils, and develop horizontal bedding.

The Silurian Longmaxi Formation is basically composed of fine clastic rocks with well-developed lamellation and horizontal bedding. The lower part is mainly composed of black shale and grayish-black silty mudstone; and the upper part is mainly composed of yellowish-green siltstone, argillaceous siltstone, and mudstone.

nodule carbonaceous mudstone of the Second Member of the Sinian Doushantuo Formation (Z_1d^2)
(Huajipo)

carbonaceous shale of the Cambrian Shuijingtuo Formation ($\epsilon_2 s$)
(Jiuqunao)

grayish-green silty mudstone of the Third Member of the Cambrian Shipai Formation ($\epsilon_2 sh^3$)
(Chayuanpo)

siliceous mudstone of the Ordovician Wufeng Formation (O_3w)
(Wangjiawan)

silty mudstone of the Silurian Longmaxi Formation (S_1l)
(Wangjiawan)

siltstone of the Jurassic Tongzhuyuan Formation (J_1t)
(well-preserved plant fossils can be seen at Guojiaba)

silty mudstone of the Silurian Luoreping Formation (S_1lr)
(Wulong; with well-preserved asymmetric ripple marks)

Figure 2-3 Types of fine clastic rocks in the fieldwork area

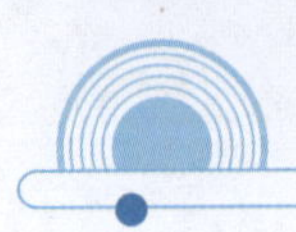

The first member of the Silurian Luoreping Formation is mainly composed of fine clastic rocks with well-developed horizontal beddings and well-preserved asymmetric ripple marks. The bottom is grayish-green silty mudstone, the middle is yellowish-green silty mudstone and shale, and the upper is grayish-green calcareous mudstone. The lower and middle members of the Silurian Shamao Formation are dominated by fine clastic rocks. The lithology is mainly yellowish-green mudstone, shale, and silty mudstone, interbedded with thin-bedded sandstone, and horizontal bedding is developed. In the middle and upper parts of the Jurassic Tongzhuyuan Formation, grayish-yellow siltstone and mudstone are developed, which are rich in plant fossils.

2.2.2 Carbonate rocks

2.2.2.1 *Limestone*

According to Dunham's carbonate classification, the common limestones in this fieldwork area mainly are: crystalline limestone without sedimentary structure; bonded limestone with sedimentary components aggregated together, mainly sponges and corals skeleton limestone; limestone with unbonded sedimentary composition.

For the third type of limestones, for the sake of summary, here is mainly based on the type and content of authigenic grain. Those with authigenic grain content less than 10% are named micritic limestone according to the Dunham classification and those with authigenic grain content over 10% are divided into bioclastic limestone, intraclast limestone, oolitic limestone, and oncolite limestone according to the main authigenic grain types.

1. Crystallized limestone

It is mainly found in the local horizon of the second member of the Sinian Dengying Formation (the Shibantan Member). The rock is relatively dense and consists of granular calcite in mosaic texture. It may be formed by recrystallization of micritic limestone (Figure 2-4).

Figure 2-4 Crystalline limestone of the Sinian Dengying Formation (Z_2dy)

2. Skeleton limestone

The reefs grown in situ are used as the skeletons. The pores inside the skeletons and among the skeletons are filled with micrite, intraclast debris, bioclastic, and sparry cements. The skeleton limestone in the fieldwork area is mainly archaeocyatha reef limestone of the Cambrian Tianheban Formation, and sponge reef limestone and coral reef limestone of the Permian Wuchiaping Formation (Figure 2-5).

archaeocyatha reef limestone of the Cambrian Tianheban Formation (ϵ_2t) (Zhouping)

sponge reef limestone of the Permian Wuchiaping Formation (P_3w) (Lianziya)

coral reef limestone of the Permian Wuchiaping Formation (P_3w)

Figure 2-5 Types of skeleton limestone in the fieldwork area

3. Bioclastic limestone

In Ordovician–Triassic carbonate rocks, bioclastic limestone is more common, mainly bioclastic granular mud-argillaceous limestone. In particular, the limestones of the Ordovician Fenxiang Formation, the Permian Maokou Formation, and the Wuchiaping Formation are rich in bioclastic content and diverse biological types (Figure 2-6).

bioclastic limestone of the Ordovician Fenxiang Formation (O_1f)
(Guiya; a large number of brachiopod biodebris can be seen on the weathering surface)

bioclastic limestone of the Ordovician Baota Formation (O_3b)
(Wulong; Sinoceras with geopetal structure)

bioclastic limestone in the Permian Maokou Formation (P_2m)
(Lianziya; with gastropods, ostracods, fuzulinid, bivalves, etc.)

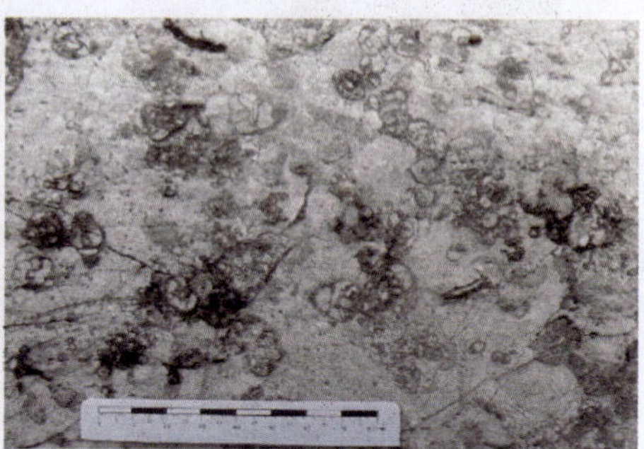

bioclastic limestone of the Permian Wuchiaping Formation (P_3w)
(Lianziya; the left are phyllophyte, sponge, and other biological debris; while the right are gastropod, sponge, ostracod, and other biological debris)

Figure 2-6 Types of bioclastic limestone in the fieldwork area

4. Intraclastic limestone

It can be seen in the partial layers of the Cambrian Tianheban Formation and the Ordovician Nanjinguan Formation. It is an intraclastic argillaceous limestone. The intraclastic grains are mostly argillaceous limestone, which is in angular–subangular shape, with a content ranging from 40% to 70%, and is a transitional support type. It is the matrix cementation and base–porous cementation type. The size of intraclastic grain is about 0.5–1 cm in the Tianheban Formation and about 0.5–4 cm in the Nanjinguan Formation (Figure 2-7).

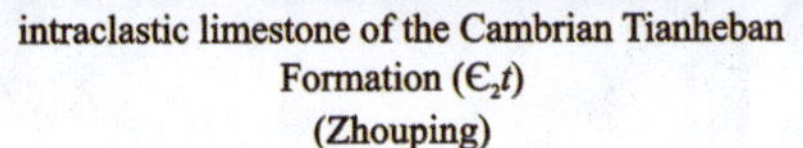

intraclastic limestone of the Cambrian Tianheban Formation (€_2t) (Zhouping)

intraclastic limestone of the Ordovician Nanjinguan Formation (O_1n) (Guiya)

Figure 2-7　Types of intraclastic limestone in the fieldwork area

5. Oncolite limestone

It can be seen in the Cambrian Tianheban Formation. The size of the oncolite is about 1–1.5 cm, and the content is about 50%. It is a transitional support type (Figure 2-8).

Figure 2-8　Oncolite limestone of the Cambrian Tianheban Formation (€_2t)
(Zhouping; part of oncolite is cut by stylolite)

6. Oolitic limestone

Oolitic limestone are found in the Cambrian Tianheban Formation, the Cambrian Shipai Formation, and other layers. The oolitic limestone in the Tianheban Formation often coexists with oncolitic limestone. The size of the oolites is about 2 mm, and the concentrical structure is clear. The oolitic content is about 40%–60%, transitional support type, matrix cementation, and base–porous cementation type. The oolitic limestone in the Fenxiang Formation has an oolitic size of about 1 mm and a content of about 80%. It is supported by grains, and it is characterized by crystallized cemented and pore cementation (Figure 2-9).

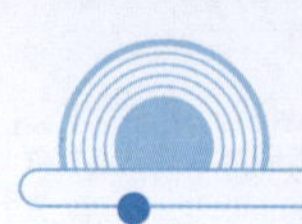

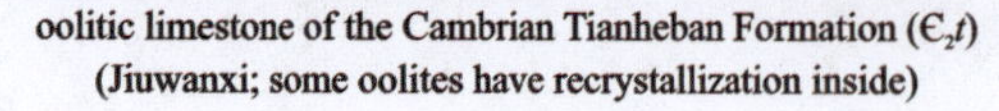

oolitic limestone of the Cambrian Tianheban Formation (Є_2t)
(Jiuwanxi; some oolites have recrystallization inside)

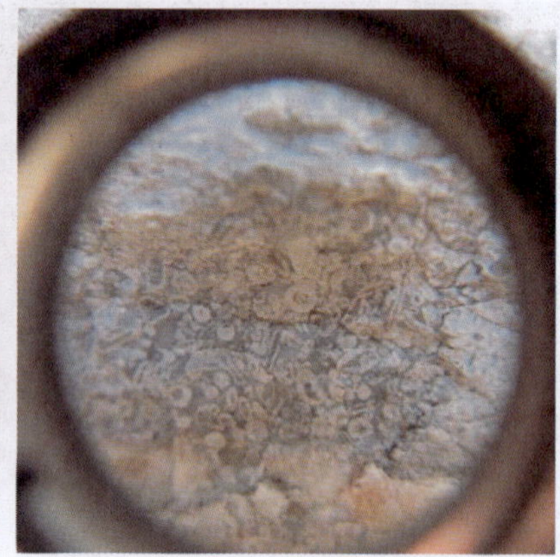

oolitic limestone of the Ordovician Fenxiang Formation (O_1f)
(Xiaojiatai, Changyang; with vision diameter of 2 cm)

Figure 2-9 Oolitic limestone of the Cambrian Tianheban Formation (Є_2t)
(Zhouping; recrystallization in some oolites)

7. Micritic limestone

It is widely distributed in the practice area, characterized by less authigenic grains, mainly the matrix of micrite, dense, and uneven fracture (Figure 2-10).

bioclastic micrite limestone of the Sinian Dengying Formation (Z_2dy)
(Wuhe; Vendotaenides can be seen)

the second member of the Cambrian Shipai Formation (Є_2sh^2) micrite limestone
(Chayuanpo)

Figure 2-10 Types of micritic limestone in the fieldwork area

2.2.2.2 *Dolomite*

The dolomites in the practice area are mainly distributed in the Neoproterozoic–the Ordovician (the Lower Paleozoic) strata and in the Upper Paleozoic–the Mesozoic strata, and mainly occur in the Triassic Jialingjiang Formation. The overall characteristic is grayish-white, with no or slow foaming when dropping dilute hydrochloric acid, and "knife-cut patterns" developing on the weathering surface. Several typical dolomites are shown in Figure 2-11.

"cap" dolomite of the first member of the Sinian Doushantuo Formation (Z_1d^1)
(Guancaiyan; with obvious "knife-cut patterns" on the weathering surface, and barite and calcareous crust at the bottom)

laminar dolomite of the first member of the Sinian Dengying Formation (Z_2dy)
(Guancaiyan; with layered"roll-up" structure)

gypsum breccia dolomite of the first member of the Sinian Dengying Formation (Z_2dy)
(Guancaiyan)

salt dome and tepee structure (Guancaiyan) in the dolomites of the first member of the Sinian Dengying Formation (Z_2dy)

dolomite of the Loushanguan Formation ($Є_3O_1l$)
(Taishangping; the layered structure can be seen in the left picture, and the stylolite structure in the right picture)

Figure 2-11 Dolomites and the sedimentary structures in the fieldwork area

2.3 Magmatic Rock and Magmatism

The intrusive rocks in the Huangling Dome area were mainly emplaced in Archean, Paleoproterozoic, and Neoproterozoic. They are predominantly composed of intermediate acid granites. These intrusive rocks provide an important window to study the Precambrian magmatism, the subduction-collision orogeny, and the crustal evolution of the Yangtze Craton. The Archean–Paleoproterozoic granitic rocks are best represented by the granitic gneiss complexes in the areas of the Dongchonghe River and the Bashan Temple, and by the Shaijiachong and Quanyitang granitic plutons. The Neoproterozoic granite is represented by the Huangling granitic batholith, which is a record of the Jinning Movement. The world-renowned Three Gorges Dam was built on the Huangling granitic batholith. The distribution of the intrusive rocks in the Huangling Dome area is shown in Figure 2-12.

2.3.1 Archean–Paleoproterozoic granitic intrusive complex

The Archean–Paleoproterozoic granitic magmatism was strongly occurred in the Huangling area. These granitic intrusions are mainly distributed in the northern part of the Huangling Dome. There are also sporadic outcrops in the Taipingxi and Dengcun areas in the south. The most representative plutons are the Archean tonalite-trondhjemite-granodiorite (TTG) rocks in the Dongchonghe River area which is in the northern part of the Huangling Dome, and the Paleoproterozoic TTG rocks in the Bashan Temple area.

2.3.1.1 *The Paleo–Mesoarchaean TTG rocks at the Dongchonghe River*

1. Geological characteristics

The Dongchonghe TTGs are distributed around the Shuiyue Temple, northwest of the Huangling Dome. This complex is in depositional contacted with the Paleoproterozoic Huanglianghe Rock Formation and the Nanhuaian–Sinian sedimentary stratum. It is also intruded by the Paleoproterozoic Quanyitang K-feldspar granite.

Inclusions are well developed in the Dongchonghe TTGs, and they can be mainly divided into two categories. One type is the wall-rock xenoliths that mainly derived from the Yemadong Rock Formation, such as amphibolite, amphibolite schists, biotite-plagioclase gneiss, and hornblendes amphibole pyroxenites. This type of xenoliths are generally

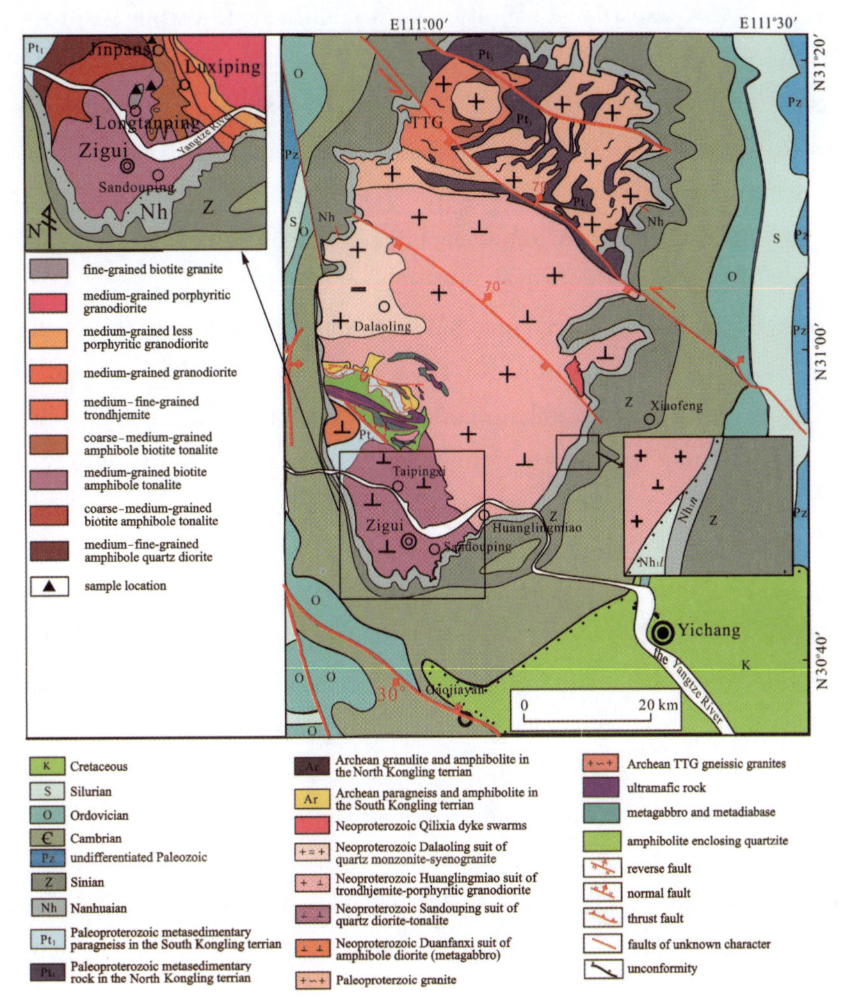

Figure 2-12 Geological map of granite distribution in the Huangling region, western Hubei (modified from Wei et al., 2012)

angular, band-shaped, strip-shaped, spherical, etc., and exhibit clear contacting boundary with the granitic gneiss. The other type is the deep-seated rock, such as mafic inclusions, which are generally not large in size. This type of inclusions, which show clear contacting boundary with their host rocks, are composed of hornblende, biotite, plagioclase, pyroxene, etc. They could be refractory residues with angular or lenticular shapes and round-edged. Parts of these enclaves show taxitic or stone sausage shapes that caused by subsequent shearing.

2. Lithological characteristics

The Dongchonghe complex is mainly composed of gneissic tonalite, gneissic granodiorite, and gneissic trondhjemite (i. e., TTG granitic gneiss), among which gneissic tonalite and trondhjemite are the majority while gneissic granodiorite is relatively rare. Gneissic tonalite and gneissic tonalite are in a transitional contact, and it is difficult to distinguish between them in the field outcrops.

(1) Gneissic tonalite. It is gray, with a granoblastic texture and a gneissose structure. It is mainly composed of plagioclase, quartz, biotite, etc., and a small amount of K-feldspars. Plagioclases, with particle sizes of 0.5–2 mm and a few grains of 2–3 mm, are classified as oligoclases, and mainly occur as xenomorphic-granular-blastic grains, with a few ones exhibit as hypidiomorphic or euhedral wide-plate columnar granular crystals. Fine and densely aggregated twin crystals are commonly developed in the plagioclases. Slight sericitic alteration can be found on the surface of the plagioclase crystals with quartz and biotite inclusions. Quartzs have xenomorphic-granular texture along feldspar grains, with a particle size ranging from 0.2 mm to 2 mm with undulatory extinction. Biotites are reddish-brown, in size of 0. 2–0. 8 mm, and show hypidiomorphic flaky texture with a few are of xenomorphic flaky texture.

(2) Gneissic granodiorite. It is blackish-gray, with a granoblastic texture and a gneissic structure. It is mainly composed of quartz, K-feldspar, and plagioclase, with about 3% muscovite in some parts. The content of the plagioclase is obviously larger than that of the K-feldspar. The K-feldspar has a xenomorphic texture, with a particle size of 1.0–2.5 mm and a perthitic texture. Mineral inclusions such as plagioclase and quartz can be seen inside of the K-feldspar. Plagioclases are granular-shaped and have a particle size of 0.8–2.5 mm, with fine-grained polysynthetic twins commonly seen. The surface of the plagioclase is commonly sericitized and metasomatized by K-feldspar. Quartz grains are granular-shaped with a particle size of 1.0–2.0 mm. Muscovites are flake-shaped with a diameter of 0.1 mm and are accompanied by a small amount of needle-columnar rutiles. It is speculated that the muscovite may be formed through the regression of biotite.

(3) Gneissic trondhjemite. Gneissic tonalites in the Dongchonghe River area contain more dark minerals than gneissic trondhjemite, thus the color of gneissic tonalite is darker than gneissic trondhjemite. Generally, the gneissic granites in the Dongchonghe River area have high contents SiO_2 and Na_2O, with $Na_2O > K_2O$, exhibit low CaO and K_2O, and are classified as metaluminous–peraluminous rocks.

3. Age of the gneissic granites in the Dongchonghe River area

Previous studies have obtained widely varied isotopic ages for the Dongchonghe River gneissic granites. However, recent researches show that the Dongchonghe gneissic granites were mainly formed around 2900–3300 Ma (Qiu et al., 2000; 焦文放等, 2009; Gao et al., 2011). Based on these studies, we consider that the Dongchonghe gneissic granites were formed in Paleo–Mesoarchean.

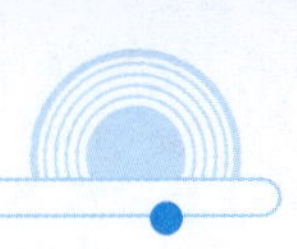

2.3.1.2 *Early Paleoproterozoic tonalite-trondhjemite-granodiorite in the Bashan Temple area*

1. Geological characteristics

The Bashansi gneissic granite, covered 57 km^2, is mainly distributed in the Wuduhe River area in the northeast of the Huangling Dome. It is in intrusive contact with the Archean Huanglianghe Formation, with its southern part intruded by the Neoproterozoic Huangling granite, and its east end in sedimentary contact with the Sinian System. TTG rocks in the Bashan Temple area covers with xenoliths developed in it. The main xenoliths in the Bashansi gneissic granite are supracrustal rocks such as amphibolite and biotite plagiogneiss, with different degrees of assimilation and contamination. The distribution of the xenoliths is generally consistent with the regional gneissosity. Relationship between inclusion and regional gneissosity is obvious. Local anatexis, which invoked by later metamorphism, can be seen in the Bashan Temple gneissic granite. The leucosomes in the migmatites are composed of coarse-grained plagioclase granite-pegmatite veins and medium–coarse-grained monzonitic pegmatite veins. They are mostly distributed along regional gneissosity but partially cut the gneissosity. In addition, the leucosomes geneally have experienced different degrees of gnetization.

2. Lithological association

The Bashansi granitic complex are mainly composed of grayish-white gneissic biotite plagiogranite and gneissic biotite monzonitic granite. These rocks have medium–fine granoblastic texture, gneissose and banded structure, with weak gneissose structure and ptygmatic folds developing in the center of the complex. They are mainly composed of plagioclase (20%–65%), quartz (20%–35%), and K-feldspar (0–30%). The majority of plagioclase crystals are xenomorphic and granular in shape with a few are of hypidiomorphic.

Most of the plagioclase grains are classified as oligoclase with a particle size of 0.3–0.5 mm. Polysynthetic twins and Carlsbad-albite twins are commonly occurred in the plagioclase crystals. The accessory minerals have complex assemblages of garnet, zircon, apatite, and pyrite, indicating the characteristics of anatexis. In terms of geochemical composition, the Bashansi granitic complex has more Na_2O content than that of the K_2O, and can be categorized as the high-alumina granite (鄂西地质大队, 1994).

3. Age of the Bashansi granitic complex

The 1/50, 000 Maoping River Regional Survey has shown that the Bashansi gneissic granites might be formed through different degrees of mixing between the basaltic and felsic

magma. They were emplaced in the Paleoproterozoic as dated by whole-rock Rb-Sr isotopic age of 2,332－2,172 Ma (姜继圣, 1986; 李福喜等, 1987).

2.3.1.3 *The Middle–Late Paleoproterozoic Shaijiachong gneissic monzonitic granite*

1. Geological characteristic

The Middle–Late Paleoproterozoic gneissic monzonitic granites are distributed in East Shaijiachong, Zhangjialaowu, Shuiyuesi and some other places. They usually outcrop as small stocks and intruded into the Dongchonghe and Bashansi granitic gneisses. Mafic inclusions are developed in some parts of these Late Paleoproterozoic gneissic granitic rocks. Gneiss folds are common in the gneissic granitic due to extensive modification. In the area of Wuduhe, the rock mass was transformed by later ductile shear to form metamorphic mylonite.

2. Lithological association

The main lithologies are hornblende-bearing biotite monzonitic gneiss, whose protolith is monzonitic granite. The rock has generally uniform fine-grained lepidoblastic texture, banded and gneissic structure. They are mainly composed of K-feldspar (25%–47%), plagioclase (20%–49%), quartz (20%–35%), biotite (3%–15%), and a small amount of magnetites. The biotite, reddish-brown, with a diameter of 0.2–0.7 mm, is intermittently distributed among the felsic minerals to form a gneissose structure. The felsic minerals have an ordered arrangement, with some parts fine-grained but recrystallized during the late stage. The plagioclase has a granoblastic texture, with a few grains preserve relict hypidiomorphic laminar–columnar texture. They are 0.2–1.0 mm in size, and generally replaced by Na-zoisites or sericite. The K-feldspars are 0.3–1.5 mm in size, develop clear lattice twin crystals under the microscope and thus can be classified as microcline. Quartz has a xenomorphic-granular texture, with a particle size of 0.1–0.8 mm, and a small amount have the sizes of 1–1.5 mm. Anatexis and potashization are common in the rocks, and some parts of the rock even have transformed to K-feldspar granitic gneiss.

3. Age of the Shaijiachong granitic gneiss

The Shaijiachong granitic gneiss intruded into the Dongchonghe and Bashansi granitic gneisses. Geochemical composition indicates that it represents the product of late-stage evolution of the calc-alkaline rocks. Therefore, the Shaijiachong granitic gneiss might emplaced later than that of the Bashansi granitic gneiss.

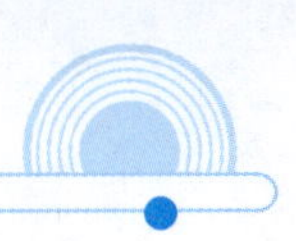

2.3.1.4 *The Late Paleoproterozoic Quanyitang K-feldspar granite*

1. Geological characteristic

The Quanyitang K-feldspar granite outcrops as a nearly equiaxed stock, with an exposed area of 21 km^2. It is distributed in the northwest of the Huangling Dome and intruded into the Archean Yemadong Rock Formation, and mainly composed of biotite K-feldspar granite. Obvious clues of assimilation and contamination can be found near the contact zone. In terms of the contact surface, the north is tilted to the south with an dip angle of 80°, and the southern tilt varies greatly, but the overall tilt is toward the south to the east, with an dip angle of 67°–84°, and a partial northward dip with an dip angle of 30°–68°. The occurrence of the contact between the granite and its surrounding rocks is mainly controlled by the schistosity of the surrounding rocks and gneiss, and the two are harmonious and consistent.

Anatexis developed at the marginal of the Quanyitang granite at Yemadong Cave, Dongchonghe River, and some other places. It is common to see that many K-feldspar veins cut the Early Archean TTG gneiss. Xenoliths are common in the rock mass, with quartzite, biotite schist, and amphibolite xenoliths commonly found in the edge. Late-stage mafic veins are commonly observed in the granite. Moreover, the crystal size of the K-feldspar granite within the Quanyitang intrusion varies obviously and clear lithofacies zoning can be observed.

2. Lithological association

The Quanyitang intrusion is dominated by biotite K-feldspar granite, followed by subordinate biotite granite, biotite monzonite, quartz syenite, syenite, etc. Among them, the quartz syenite, monzonite, and syenite are mainly distributed in the southern part of the intrusion (Table 2-2).

The rock is brick-red, with a porphyritic texture and a massive structure. The main minerals are K-feldspar (65%–70%), quartz (20%–25%), biotite (<5%), plagioclase (<5%). Magnetite, apatite, and zircon are the main accessory minerals (< 1%). The K-feldspar, mainly classified as microcline and a small amount of perthites, has an obvious graphic texture, with a particle size up to 5 mm. Quartz has a hypidiomorphic–idiomorphic texture, with a particle size of 1–2 mm.

3. Age of Quanyitang K-feldspar granite

U-Pb dating and geochemical composition have revealed that the Quanyitang intrusion emplaced at the late Paleoproterozoic (ca. 1850 Ma) and is classified as A-type granite. It was formed by partial melting of the deep Archean crust under a post-orogenic extensional tectonic regime in the Paleoproterozoic (熊庆等, 2008; Peng et al., 2012).

Table 2-2 Rock characteristics of the Quanyitang K-feldspar granite in the core of the Huangling Dome

Rock Type	Main Mineral Content/%				Texture and Structure
	K-feldspar	Plagioclase	Quartz	Biotite	
Biotite-bearing alkali granite	56–64	5–10	28–32	3–6	mainly granitic and metasomatic texture, followed by porphyritic-like texture, microscopic graphic texture, and graphic-like texture; massive structure
Biotite K-feldspar granite	44–48	20–28	23–30	1–4	granitic texture, metasomatic texture; massive structure
Biotite monzonitic granite	27–47	25–30	25–36	3–4	mainly granitic texture, followed by metasomatic texture, microscopic graphic texture, graphic-like texture, fragmented texture, and porphyritic-like texture; massive structure
Biotite quartz syenite	64–67	8–10	18–20	5–7	granitic texture; massive structure
Biotite quartz monzonite	28–40	32–50	10–15	3–5	hypidiomorphic–xenomorphic granular texture, metasomatic texture (granitic texture)
Syenite	70–80	1	2–3	1–2	metasomatic texture; massive structure

2.3.2 The Neoproterozoic granitic intrusive complexes

The Neoproterozoic granitic intrusive complex refer to the Neoproterozoic Huangling granitic complex which is also known as the Huangling granitic batholith or the Huangling composite granitic body that mainly distributed in the south of the Huangling Dome. Integrate the 1/50,000 Liantuo and Sandouping Regional Geological Mapping accomplished by the Wuhan Geological Survey Center (2012), the Division Scheme of the Neoproterozoic granitic intrusive complex of Wei et al. (2012) and Ma et al. (2002), and the 1/250,000 Jingmen Regional Survey (2006), etc., the Neoproterozoic Huangling granite complex can be divided into four assemblages corresponding to four stages of magmatic activitives from early to late (see Figure 2-12 for details).

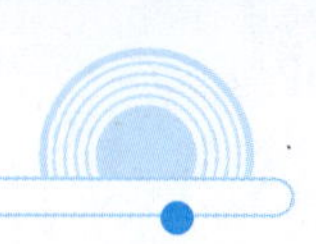

2.3.2.1 *Stage I: the Neoproterozoic neutral–basic intrusive rocks*

1. The Yazikou medium–fine-grained diorite ($Pt_3^1\delta$)

1) Geological features

The Yazikou intrusion intruded into the Xiaoyucun Formation and was partially cut by the Huangling unit, in which a large number of xenoliths of the Yazikou unit are found.

2) Petrographical characteristics

This intrusion is mainly composed of medium–fine-grained meta-diorite and shows piebald structure due to the heterogeneous distribution of partially dark minerals in the pattern of flowers; hypersthene and augite remnants are occasionally seen in the rock. The main mineral contents of the rocks are as follows: plagioclase (77%–78%), amphibole (20%–21%), biotite (1%–2%), and pyroxene (1%). There are a few kinds of accessory minerals, dominated by magnetite, followed by pyrite and apatite. The rock mass contains plagioclase amphibolite, amphibolite and biotite plagioclase gneiss inclusion, etc. The characteristics of plagioclase amphibolite and plagioclase gneiss enclaves are similar to the Kongling Rock Group. The enclaves have a gradual relationship with the surrounding rock, and such enclaves should be the remains of deep magma melting. According to the geological fact that the Yazikou rock mass is cut by the Huangling superunit, it is speculated that the formation age of the rock mass should be greater than 860 Ma.

2. The Zhaibao fine–medium-grained diorite body ($Pt_3^1\delta z$)

1) Geological features

The Zhaibao rock mass intrudes into the Yazikou unit, and the contact interface is clear in the form of a bay, which is inclined inward. A relatively dense foliation zone with a width of about one meter can be seen in the inner contact zone. The northwest is covered by the sedimentary contact of the Liantuo sandstone.

2) Petrographical characteristics

The rock mass is mainly composed of fine–medium-grained diorite, and the main mineral contents are plagioclase (59%–60%), amphibole (32%–33%), pyroxene (5%–6%), and biotite (1%–2%). There are a few kinds of auxiliary minerals in the rocks, dominated by magnetite, followed by pyrite, apatite, etc. There are few inclusions in the rock mass, which are mainly amphibolite, and amphibolite is distributed near the inner contact zone. According to the geological contact relationship, the formation age of this rock mass should be slightly later than that of the Yazikou medium–fine-grained diorite.

2.3.2.2 *Stage Ⅱ: the Neoproterozoic neutral–acidic intrusive rocks (TTG?)*

The second period of Neoproterozoic neutral–acidic intrusive rock assemblage is located in the southwest of the Huangling Dome, distributed in Sandouping and Huangjiachong areas, generally distributed in the north–northwest direction, and the northwestern side intruded into the Miaowan Formation (i. e., Miaowan ophiolitic mélange), the southern part is unconformably overlain by the Nanhuaian Liantuo Formation, and the eastern side is intruded by the Huangling superunit. The assemblage is characterized by quartz diorite to tonalite, with fine–coarse-grained heterogranular structure, massive structure. The main minerals include plagioklase, amphibole, quartz, biotite, etc., which are subaluminous calc-alkaline to acidic rocks.

Micrograined enclaves are well developed in the rocks. According to the characteristics of lithology, mineral composition, structure, enclave, and contact relationship, the enclaves can be divided into four magma-intruded rock bodies (units): Zhongba medium–fine-grained quartz diorite ($Pt_3^2\delta o$), Taipingxi medium–coarse-grained quartz diorite ($Pt_3^2\delta o$), Sandouping tonalite ($Pt_3^2\gamma o\beta$), and Jinpansi tonalite ($Pt_3^2\gamma o\beta$).

1. The Zhongba medium–fine-grained quartz diorite ($Pt_3^2\delta o$)

1) Geological features

The Zhongba unit (pluton) outcropped in an arc-like shape which is nearly northsouth–northeast distributed. Its western side intrudes into the Kongling Rock Group, the southern part is unconformably overlain by the Liantuo Formation, the eastern side shows parallel intrusive unconformable contact with the Taipingxi unit, and the southeastern side is cut obliquely by the Sandouping unit.

2) Petrographical characteristics

The main lithology of the Zhongba pluton is medium–fine-grained quartz diorite. The main minerals of the rock are plagioclase (54%–55%), hornblende (32%–33%), quartz (10%–11%), and biotite (2%–3%). There are a few types of accessory minerals in the rock, with magnetite being dominant, and a small amount of zircons, apatites, pyrites, etc.

Many types of xenoliths are developed in the Zhongba unit, including micro-fine-grained diorite (porphyrite), amphibolite, (hornblende) biotite plagioclase gneiss, etc. In addition, quartz diorite and diabase porphyrite xenoliths are also seen. The amphibolite and biotite plagioclase gneiss xenoliths are mostly produced near the inner contact zone between the pluton and the Kongling Rock Group, and have similar petrographical features to those of the metamorphic rocks in the Kongling Rock Group. The xenoliths are single isolated occurred or together appeared in linear distribution, and in truncated or diffuse contact with the host rocks. Zoned biotite edges are occasionally developed in some of the xenoliths.

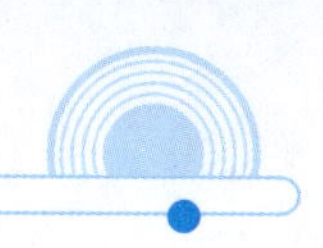

According to the field contact relationship, the Zhongba fine-grained quartz diorite should be formed earlier than the Sandouping tonalite, that is, earlier than 860 Ma, but later than the Zhaibao fine–medium-grained gabbro that emplaced in the first period of the Neoproterozoic.

2. The Taipingxi medium–coarse-grained quartz diorite ($Pt_3^2\delta o$)

1) Geological features

The Taipingxi medium–coarse-grained quartz diorite is distributed in a band form which is nearly northsouth–north–northeast exhibited. Its southeastern part was cut by the Sandouping unit, and the northern side intruded into the Kongling Rock Group.

2) Petrographical characteristics

The main lithology of the Taipingxi pluton is medium–coarse-grained quartz diorite. The main minerals of the rock are plagioclase (64%–66%), quartz (14%–16%), hornblende (11%–13%), and biotite (5%–6%). There are a few types of accessory minerals, with magnetite being dominant, and many apatite and allanite.

Xenoliths are widely developed in the Taipingxi pluton. They are mainly of diorite porphyrite, in the form of long strips to lens with smooth shape. Most of the xenoliths are densely together occurred in band distribution which generally ranges from 3 m to 5 m in width and is parallel to the trend of local foliation. The minerals of the diorite porphyrite xenoliths in the Taipingxi pluton are similar to those in the diorite (porphyrite) xenoliths in the Zhongba medium–fine-grained quartz diorite, except the occurrence of plagioclase phenocryst (5%–8%).

According to the field contact relationship, the Taipingxi medium–coarse-grained quartz diorite should be formed earlier than the Sandouping tonalite, that is, more than 860 Ma, but later than the Zhongba medium–fine-grained quartz diorite.

3. The Sandouping tonalite ($Pt_3^2\gamma o\beta$)

1) Geological features

The Sandouping unit is the main part of the second stage of the Neoproterozoic intrusive complexes. It is nearly north-south distributed around the Sandouping and Wangliangchuya areas. Regional survey reveals that the northern part of the Sandouping uint intruded into the Mesoproterozoic Xiaoyucun Formation of the Kongling Rock Group and the Meso–Neoproterozoic Miaowan Formation, with the southern section was unconformably overlain by the Nanhuaian Liantuo Formation, and the eastern domain cut by the Jinpansi tonalite ($Pt_3^2\gamma o\beta$) and the Luxiping plagiogranite (trondhjemite) ($Pt_3^2\gamma o$).

2) Petrographical characteristics

The main lithology of the Sandouping unit is medium-grained biotite hornblende tonalite.

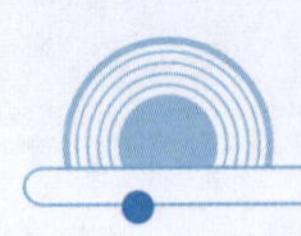

The weathered surface of the rock is greyish-brown, and the fresh surface is dark gray mottled with black and white. The rock mainly shows medium-grain texture, with felsic mineral grains ranging from 2 mm to 4 mm and a small amount even up to 5 mm, and massive structure. The main minerals are plagioclase (55%–65%), quartz (10%–18%), biotite (12%–20%), hornblende (5%–10%), etc. Commonly seen accessory minerals are magnetite, followed by subordinated apatite, ilmenite, allanite, zircon, etc. The colors of zircons are relatively miscellaneous, mainly of rose and light yellow. Geochemical data indicate that the Sandouping pluton can be classified as peraluminous calc-alkaline granite. Xenoliths, including diorite (porphyrite), dark diorite, and amphibolite, are commonly developed in the Sandouping unit.

The Sandouping tonalite intruded into the Meso–Neoproterozoic Miaowan Formation ($Pt_{2-3}m$), but was in turn intruded by the third period emplaced Huangling granite during the Neoproterozoic. Medium-grained hornblende biotite tonalite from the Sandouping unit gives zircon SHRIMP U-Pb age of (863±9) Ma (Wei et al., 2012).

4. The Jinpansi tonalite ($Pt_3^2\gamma o\beta$)

1) Geological features

The Jinpansi unit is distributed in a northnorthwest direction. The western part of this pluton is in surge contact with the Sandouping tonalite, the southern part is unconformably overlain by the Nanhuaian sediments, and the eastern domain is intruded by the Luxiping plagiogranite (trondhjemite).

2) Petrographical characteristics

The main lithology of the Jinpansi pluton is medium–coarse-grained hornblende biotite tonalite. The rock has medium–coarse-grained texture and massive structure. Its main minerals are plagioclase (55%–62%), quartz (12%–20%), biotite (12%–18%), and hornblende (7%–12%). The plagioclase is mostly hypidio-banded in shape, with the particle size of 2–5 mm. The Biotite is mostly scaly, page-like, and aggregated, with slice diameter mainly of 2–5 mm and some ones up to 7–10 mm. The hornblende exhibits as hypidio-long column, with length of 3–6 mm and a small amount of up to 8 cm. Commonly seen accessory minerals are magnetite, apatite, zircon, allanite, etc. Geochemical data indicate that the Jinpansi pluton belongs to aluminous calc-alkaline granite. The common seen xenoliths in the Jinpansi pluton include diorite porphyrite and amphibolite. They have smooth shapes and mostly occur as single ones. Circled biotite can be occasionally seen around the edges of these xenoliths.

The Jinpansi tonalite intruded into the Meso–Neoproterozoic Miaowan Formation ($Pt_{2-3}m$), but was intruded by Huangling granite of third period of Neoproterozoic. The isotopic diagenetic age of the zircon SHRIMP U-Pb obtained from the coarse–medium-grained biotite hornblende diorite is (842±10) Ma (Wei et al., 2012).

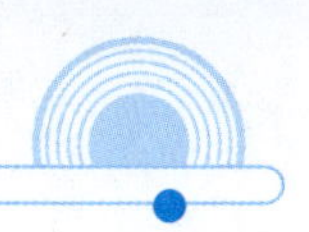

2.3.2.3 *Stage Ⅲ: the Neoproterozoic intrusive rocks*

The third stage of Neoproterozoic intrusive rocks constitute as the main part of the Huangling granite batholith. They are distributed in the Yingzizui, Neikou, Guchengping, and other places. The west side of this episode of rocks intruded into the second stage of Neoproterozoic intermediate–acid intrusive rocks, and the south end is unconformably covered by the Nanhuaian Liantuo Formation. Generally, the third stage of intrusive rocks have fine–coarse medium-grained or continuous unequal grained structure, massive structure, with simple types of xenoliths that sporadically occurred. According to mineral assemblages, texture, structure, and filed contact relationship, the third stage of Neoproterozoic intrusions can be further divided into four units, including the Luxiping plagioclase (trondhjemite) granite ($Pt_3^3\gamma o$), the Yingzizui granodiorite ($Pt_3^3\gamma\delta$), the Neikou porphyric granodiorite (monzogranite) ($Pt_3^3\pi\gamma\delta$), and the Maopingtuo porphyritic granodiorite (monzogranite) ($Pt_3^3\pi\gamma\delta$).

1. The Luxiping plagioclase (trondhjemite) granite ($Pt_3^3\gamma o$)

1) Geological features

The Luxiping unit is distributed in a NNW and NW trending belt. The intrusion intruded into the Jinpansi coarse–medium tonalite in an oblique style, and also intruded into the Neoproterozoic Miaowan ophiolitic mélange. The east side of the Luxiping unit mainly show surge, and some local areas pulsating contact relationship with the Yingzizui medium-grained granodiorite. In the Gehouping Village, the Luxiping unit outcrops in nearly north–south belt, with its northwest side in intrusive contact with the medium grained granodiorite, and the rest part being unconformably covered by the Nanhuaian or Sinian sedimentary system.

2) Petrographical characteristics

The Luxiping unit is mainly medium–fine-grained plagioclase (trondhjemite) granite, with part of tonalite. The weathered surface of the rock is grayish-yellow, and the fresh surface is gray. The rock has medium–fine granitic texture and massive structure, with most mineral grain size of 1–2.5 mm. The main minerals are plagioclase (64%–68%, in the shape of xenomorphic-hypidiomorphic lath, with the development of albite twins, occasionally with Carlsbad-albite compound twins, and zoned texture), quartz (24%–30%); biotite (4%–8%, mostly in the form of flakes, a small amount of leaf sheet-like and orientated), amphibole (1%–3%, in the form of needle column), and K-feldspar (2%–5%). The accessory minerals include magnetite, a small amount of monazites, garnets, zircons, etc. Zircon is rosy–light-rosy and well developed with ring structure. Geochemical data shows that the Luxiping unit can be classified as peraluminum calc-alkaline granite. Medium–coarse-grained or porphyritic biotite-quartz-diorite and medium–fine-grained quartz-diorite xenoliths can be occasionally found in the Luxiping unit. Plagioclase amphibolite and biotite plagioclase gneiss xenoliths are also found near the contact area between the Luxiping intrusion and the Kongling Rock Group.

The Luxiping plagioclase (trondhjemite) granite intruded into the Neoproterozoic Miaowan ophiolitic mélange and medium–fine-grained tonalite, but was instead intruded by the Yingzizui medium grained granodiorite. Zircon SHRIMP U-Pb dating obtained from the Luxiping medium–fine-grained plagioclase (trondhjemite) granite gives an age of (852 ± 12) Ma (Wei et al., 2012).

2. The Yingzizui medium-grained granodiorite ($Pt_3^3\gamma\delta$)

1) Geological features

This intrusion is distributed around Yingzizui and outcrops in an annular shape. The east side of the Yingzizui intrusion is composed by six small plutons all of which are distributed in the northwest direction; while the west side consists of a large pluton that occurs as strips in the northwest direction. This unit intruded into the Luxiping medium–fine-grained plagioclase (trondhjemite) granite, and in turn was intruded by the late Maopingtuo medium-grained (porphyritic) porphyritic granodiorite, and was pulsated by the Neikou medium-grained porphyritic granodiorite.

2) Petrographical characteristics

The Yingzizui unit is mainly medium-grained granodiorite with medium-grained texture. The mineral grain size of the rock varys from 2 mm to 5 mm, and mostly of 3 mm. The rock mainly consists of plagioclase (50%–55%), quartz (25%–30%), K-feldspar (8%–15%), and biotite (4%–5%). The plagioclase is yellowish-brown in color, hypidiomorphic lamellar strip in shape, with polysynthetic twins and occasionally Carlsbad-albite compound twins developed. In some rocks, the surface of the plagioclase crystal is turbidized, where clayization and sericinization can be observed. Some parts of the plagioclase grains are metasomatized by muscovite veins. The quartz is xenomorphic granular, and shows wavy extinction and recrystallization in some parts due to tectonic modification. The K-feldspar is xenomorphic granular–hypidiomorphic tabular in shape, generally with lattice twin crystals and occasionally of perthite (orthostriped feldspar), and unevenly distributed in rock. The biotite is scaly with a few grains in pagelike shape and displays pleochroism of light yellow–dark brown. In the rocks near the Nantuo Village, it can be seen that some of biotites are metasomatized by muscovites, and a small amount are metasomatized by chlorites. The accessory minerals are mainly magnetite, followed by apatite, zircon, and allanite. Zircons are of various colors, mainly light rose and light

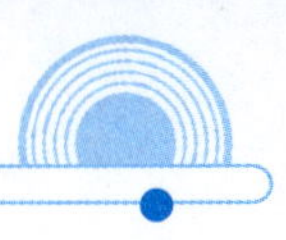

yellow, followed by lavender. Diorite porphyrite and mafic coarse-grained dioritic xenoliths are common in the Yingzuizi intrusion. Porphyry biotite quartz diorite and medium–fine-grained biotite tonalitic xenoliths are occasionally seen, and plagioclase amphibolite and gneiss xenoliths can be seen at the contact zone with the Kongling Rock Group. Geochemical data show that the Yingzizui intrusion belongs to peraluminum calc-alkaline granite.

The Yingzizui unit is in surge contact with the Luxiping medium-grained plagioclase (trondhjemite) granite and the Maopingtuo medium-grained porphyric granodiorite, and is pulsating intruded by the Neikou medium granitic porphyritic granodiorite. Zircon SHRIMP U-Pb dating conducted on the granodiorite in the Yingzizui unit gives an age of (850±4) Ma (Wei et al., 2012).

3. The Maopingtuo phenocryst-bearing medium–grained granodiorite ($Pt_3^3\pi\gamma\delta$)

1) Geological features

The Maopingtuo phenocryst-bearing medium-grained granodiorite unit is distributed around Maopingtuo near the Letian Creek. The unit is in inrush contact with the Yingzizui medium-grained granodiorite and the Neikou medium-grained porphyritic granodiorite.

2) Petrographical characteristics

The Maopingtuo unit is mainly phenocryst-bearing medium-grained granodiorite. The weathered surface of the rock is grayish-yellow, while the fresh surface is light gray. The mineral grain size of the rock varys from 2 mm to 5 mm. The main minerals are plagioclase (55%–60%), quartz (28%–35%), K-feldspar (3%–8%), and a small amount of biotites (3%–5%). The accessory minerals are magnetite mainly, while the content of the other accessory minerals is rather low. The rock has porphyritic texture and massive structure. The phenocrysts are mainly aggregated quartzes and a small amount of plagioclases. The K-feldspar phenocrysts are rather rare. The low K-feldspar content makes the rock close to the light color tonalite.

The Maopingtuo phenocryst-bearing medium-grained granodiorite unit is distinguished from the Yingzizui medium-grained granodiorite by containing plagioclase and quartz phenocrysts. When compared with the Neikou unit, it found that the Neikou medium-grained porphyric monzonitic granite is dominated by more than 10% K-feldspar phenocrysts which are larger, while the Maopingtuo phenocryst bearing medium-grained granodiorite unit contains mainly of aggregated quartz phenocrysts instead of K-feldspar phenocrysts. Geochemical data show that the Maopingtuo unit belongs to peraluminum calc-alkaline granite. Dominated dioritic porphyrite and mafic coarse-grained dioritic enclaves, and occasionally porphyritic biotite tonalitic and medium–fine-grained biotite tonalitic xenoliths are developed in the Maopingtuo unit; in addition, amphibolite and gneiss xenoliths can also be found within the contact zone between the intrusion and the Kongling Rock Group.

The Maopingtuo phenocryst-bearing medium-grained granodiorite unit intruded into the

Neoproterozoic Miaowan ophiolitic mélange and the fine–medium-grained tonalite, and also in inrush contact with the Yingzizui and Neikou units. Zircon SHRIMP U-Pb dating obtained in the Maopingtuo phenocryst-bearing medium-grained granodiorite gives an age of (844±11) Ma (Wei et al., 2012).

4. The Neikou medium-grained porphyritic granodiorite ($Pt_3^3\pi\gamma\delta$)

1) Geological features

The Neikou unit is mainly distributed around the Letian Greek, Guchengping, and Zhongguzhai areas. It is in inrush contact with the Maopingtuo unit, and in pulsating intrusive contact with the Zongxifang intrusion.

2) Petrographical characteristics

The Niekou unit is mainly medium-grained porphyritic biotite granodiorite, while in some places the rock has relatively high K-feldspar content and can be named as monzogranite. The rock has porphyritic texture, massive structure, with a mineral grain size of 2–5 mm. The weathered surface of the rock is grayish-yellow, while the fresh surface is light gray. The main minerals are plagioclase (52%–55%), quartz (28%–33%), K-feldspar (10%–20%) and a small amount of biotites (3%–5%). Zoned texture is common in the K-feldspar. The accessory minerals are mainly magnetite, with a small amount of allanites, titanites, zircons, etc. Porphyritic biotite diorite, porphyritic biotite quartz diorite, dioritic porphyrite, biotite schist and other xenoliths are found sporadically occurred in the rock. Most of the xenoliths are subcircular to subangular in shapes, while the medium–fine-grained biotite tonalitic xenoliths occur in strips, and are in truncated contact with their country rocks. Geochemical data show that the Neikou unit belongs to peraluminum calc-alkaline granite.

The Neikou medium grained porphyritic granodiorite unit intruded into the Yingzizui medium grained granodiorite, and at some areas, it is in pulsating intrusive contact with the Maopingtuo unit. Zircon SHRIMP U-Pb dating obtained from the medium-grained porphyritic biotite granodiorite in the Neikou unit gives an age of (835±14) Ma (Wei et al., 2012).

2.3.2.4 *Stage IV: Neoproterozoic intrusive rock*

This stage of rocks are mainly distributed around the Dalaoling Forest Farm in the northwest of the Huangling granitic batholith, including four magmatic intrusion units. The western part is unconformably covered by the Sinian System. The northern, eastern and southern sides intruded into the third stage of Neoproterozoic intrusions and the Archean Kongling Rock Group, with a formation age of 795 Ma (Ling et al., 2006).

1. The Fenghuangping monzodiorite ($Pt_3^4\eta\delta$)

This unit is distributed on the northeast edge of the Dalaoling superunit and generally arc-shaped. The rock has high color index, medium-grained texture, massive structure (partly banded), and slightly planar structure.

2. The Tianjiaping porphyaceous hornblende biotite monzogranite ($Pt_3^4\pi\eta\gamma$)

This unit is near east–west distributed, and different from the Gujiangping unit with a large amount of coarse phenocrysts of K-feldspar and obvious hornblendes. The contact relationship between the Tianjiaping unit and the Gujiangping unit has not been identified. Comparatively, the Tianjiaping unit has higher color index and hornblende but lower SiO_2 contents relative to the Gujiangping unit. Thus, the Tianjiaping unit should be formed earlier than the Gujiangping unit according to the magmatic evolution law.

3. The Gujiangping monzogranite ($Pt_3^4\eta\gamma$)

It is the largest intrusion of the Dalaoling superunit, mainly distributed at Zhiziguai, Dalaoling Forest Farm, Tianzhushan Mountain, Changchong, and its west areas. The Gujiangping unit is obliquely cut into the Fenghuangping unit, and sometimes the two units show gradual transition relationship.

4. The Mahuagou garnet-bearing monzogranite ($Pt_3^4\eta\gamma$)

This unit includes the Mahuagou, Shaping, Longtansi, and some other intrusions, as well as many uncounted vein-like small intrusions. In addition, this unit intruded into the Huangling and Sandouping superunits, respectively, but has not been found to be in contact with the other units of the Dalaoling superunit. According to the texture and mineral composition, here we temporarily consider that the Mahuagou unit is the latest intrusion of the Dalaoling superunit.

2.3.3 The Neoproterozoic mafic–intermediate dyke groups

The Neoproterozoic mafic–intermediate dyke group is mainly distributed at Xiaofeng that on the east side of the core of the Huangling Dome and named as the Xiaofeng suite or the Qilixia dyke group ($Pt_3\delta\mu$-$\gamma\pi q$). Single dyke of the group is small in scale, but abundant in number and rather varies in lithology. In addition, various veins are widely developed in the mafic–intermediate dykes, with the strikes mostly of NE30°–70°. In the north, west, and south of the Huangling Dome, the Neoproterozoic mafic–intermediate dykes generally intruded into the Luxiping and Neikou units, showing hyperkinetic contacts. These dyke groups are composed by a large number of NE-trending steep dykes (veins). Single dyke is generally 1–10 m wide and 30–70 m long along the strike. Most of the dykes dip to the southeast and a few to the northwest. The formation age of these dykes is 806–797 Ma (Zhang et al., 2008).

The Qilixia dyke group has complex lithology, mainly including fine-grained diorite, dioritic porphyrite, quartz dioritic porphyrite, quartz monzodioritic porphyrite, plagiogranitic porphyry, etc. These dykes have clear and distinct contact boundaries with the surrounding rocks, and the intrusive relationship between them is that plagiogranitic porphyry dyke intruded into the surrounding rocks, dioritic porphyrite dyke intruded into the fine-grained diorite, quartz dioritic porphyrite dyke intruded into the dioritic porphyrite vein, quartz monzodioritic porphyrite dyke intruded into the dioritic porphyrite vein, etc.

The emplacement sequence of the Qilixia dyke group is fine grained diorite → dioritic porphyrite → quartz dioritic porphyrite → quartz monzodioritic porphyrite → granitic porphyry. In addition, a small number of microdioritic dykes and diabase (porphyrite) veins are randomly distributed, with the same occurrence as the above-mentioned dykes, and obviously cut through the above dykes. Dark enclaves also can be seen in the plagiogranitic porphyry dykes, with various shapes, such as round, leaf-shaped, irregular, etc. Generally speaking, the larger the enclaves, the more irregular the shape.

(1) Fine-grained diorite dyke. This type of dyke is often intruded by dioritic porphyrite dykes, with clear boundary between each other. 1–2 mm baked rim can be seen at the edge of the fine-grained dioritic dyke, and the occurrence of the contact surface is 300°∠79 °. The dyke is gray, fine-grained texture, massive structure, with main minerals of plagioclase, amphibole, biotite, and a small amount of quartzs. Plagioclase is xenomorphic, hypidiomorphic-granular, or lath-shaped, with a particle size of 0.5–2 mm; amphibole is in short column shape with particle size of 1–2 mm. The biotite is in the form of fine scales. The accessory minerals are mainly magnetite and sphene.

(2) Dioritic porphyrite dyke. It is the main type of dyke, often intruded into fine-grained dioritic dyke, and intruded by quartz monzodioritic porphyrite dyke. The rock is dark gray, porphyritic texture, massive structure, and its main mineral composition is shown in Table 2-4. The phenocrysts are mainly composed of plagioclase and amphibole, with a small amount of biotites. Amphibole is idiomorphic column; plagioclase is mostly the idiomorphic–lath-shaped, and a few are round-shaped due to corrosion, with the maximum particle size of 3–5 mm. The matrix is mainly of cryptocrystalline, accounting for about 70% of the total rock. The accessory minerals are magnetite, apatite, and zircon.

(3) Quartz dioritic porphyrite dyke. The dyke is gray in color, with porphyritic texture and massive structure, and its main minerals are shown in Table 2-3. The phenocrysts are mainly of plagioclase, euhedral–hypidiomorphic lath-shaped, and can be classified as andesine, with a particle size of 0.4 mm × 10 mm–1 mm × 4 mm. Zoned texture and

polysynthetic twins can be seen in the plagioclase grains, some of which also show sericitization and epidotization. Matrix in the rock is of fine-grained texture. The accessory minerals are apatite, zircon, epidote, sphene, etc.

Table 2-3 Mineral contents of each rock types in the Qilixia dyke swarm ($Pt_3\delta\mu$ -$\gamma\pi$ q)

(According to 1/50,000 regional survey report of Liantuo, Fenxiang, Sandouping, and Yichang, 2012)

Name	Lithology	Main Mineral Content/%				
		Orthoclase	Anorthose	Quartz	Biotite	Amphibole/Hornblende
The Qilixia cluster (dyke)	plagiogranitic	2–3	60	30	1–2	
	quartz monzodioritic porphyrite	10 (phenocryst)	12 (phenocryst)			
	quartz dioritic porphyrite	5	30–40	15–20	3–5	5–10
	dioritic porphyrite		60	5	10	10–20
	fine-grained diorite		60	5	4–5	20

(4) Quartz monzodioritic porphyrite dyke. The dyke often contains fine-grained dioritic and dioritic porphyritic veins, and is purplish-red in color, with porphyritic texture and massive structure. The phenocrysts in the rocks are mainly plagioclase and K-feldspar, which are both euhedral lath-shaped, with a particle size of 3 mm×2 mm. The plagioclase develops Carlsbad-albite compound twins, while the K-feldspar shows Casbah twins. Matrix in the rock shows fine-grained texture. Accessory minerals mainly are magnetite and apatite.

(5) Plagioclase granitic porphyric dyke. It often intruded into quartz monzodioritic porphyric dyke. The rock is light red–purplish red in color, with porphyritic texture and massive structure, and its main minerals are shown in Table 2-4. Phenocrysts are plagioclase, idiomorphic lath-shaped, with a few grains are round due to corrosion. Generally, the plagioclase phenocrysts have a particle size of 4 mm×3 mm–8 mm×5 mm,

and are developed with polysynthetic twins and zoned texture. The matrix is mainly composed of quartz, euhedral or not regular granular, with a particle size of 2 mm×1.5 mm. The matrix shows exocrystalline-cryptocrystalline texture, and is composed of quartz, plagioclase, and biotite. Accessory minerals of the rock are apatite, rutile, zircon, etc.

The Qilixia dyke swarm in the core area of the Huangling Dome has an obvious preferred orientation. Its spatial distribution is generally northeast, with a steep contact interface with the surrounding rock. Magmatic intrusion structures such as condensation edge, which are obviously controlled by two groups of regional northeast and northwest faults, also can be observed. These suggest that the Qilixia dyke swarm emplaced under lithospheric extension. Thus, the period of the emplacement of the Qilixia dyke swarm has entered into the post orogenic extensional tectonic regime that accompanied by obvious uplift.

2.3.4 The Meso-Neoproterozoic mafic-ultramafic rocks

In the 1960s and 1970s, the Exi Geological Team of Hubei Province, Yichang Institute of Geology and Mineral Resources, and other units carried out geological exploration, prospecting, and researching on chromite in the metamafic-ultramafic rocks distributed in Taipingxi and Dengcun areas in the south of the Huangling Dome (1/50, 000 regional geological survey and mapping), and named them the Miaowan Formation. Inrecent years, Peng Songbai et al. (2010) and Peng et al. (2012) proposed a set of meta-mafic-ultramafic rocks based on detailed field geological surveys, petrography, geochemistry, and structural deformation characteristics. The meta-mafic-ultramafic rock is actually a new understanding of a set of Neoproterozoic ophiolite fragments and it is named Miaowan ophiolite.

The Meso-Neoproterozoic mafic-ultramafic rocks are mainly distributed in the Dengcun and Xiaoxikou areas, generally distributed in a NWW trending band, and are also the largest ultramafic bodies exposed in the central southern China region. Meta ultramafic rocks are continuously exposed for a maximum length of 13 km and a width of nearly 2 km. Layered meta mafic rocks and metasedimentary rocks are distributed on both sides of the meta ultramafic rocks. The meta ultramafic rocks are mainly serpentinite, serpentinized dunite, and pyroxene peridotite. Meta mafic rocks are mainly layered fine-grained amphibolite and layered massive metagabbro bodies, dykes and diabase veins are distributed in layered fine-grained amphibolite, serpentinized dunite, and harzburgite (Figure 2-13). In addition, the meta ultramafic and mafic rocks are closely associated with a small amount of lenticular and thin-bedded marbles, quartzites, and other metasedimentary rocks.

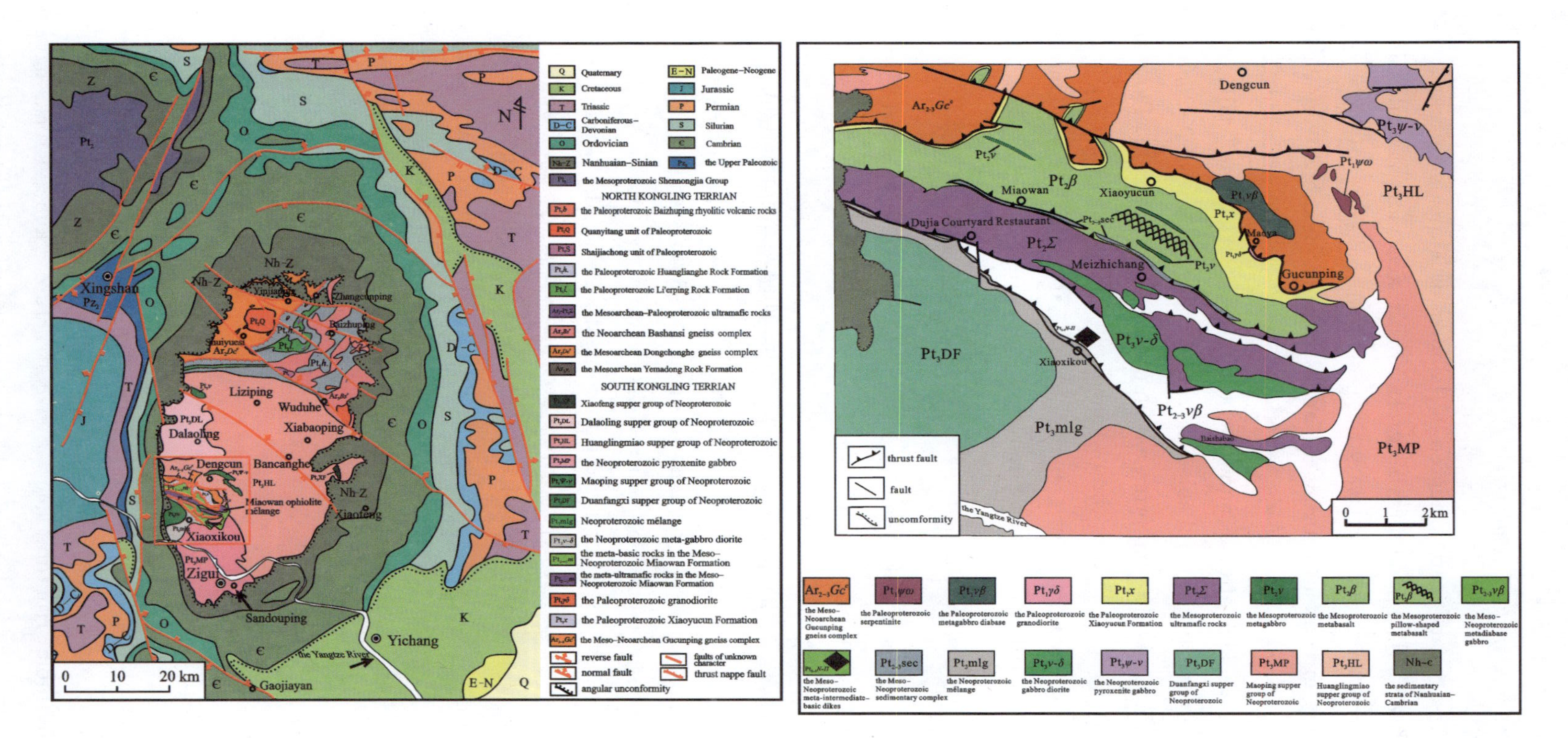

Figure 2-13 Geological structure sketch in the southern Huangling Dome (modified from 彭松柏等, 2010 and Peng et al., 2012)

2.3.4.1 *Serpentinized wehrlite*

The serpentinized wehrlite are produced in the form of lenticular blocks and slices. The rocks are dark grayish-black and grayish-green, with xenomorphic–subhedral columnar texture, net vein structure, and massive structure. The rocks are strongly serpentinized, with the minerals are directionally arranged, and mylonitic foliations were developed. The main minerals are pyroxene (45%–50%), olivine (35%–45%), hornblende (3%–5%), and magnetite (1%–2%). The altered minerals are mainly serpentine, talc, and chlorite. Olivine is hypidiomorphic to euhedral columnar, with a particle size of 3–5 mm, and mostly replaced by serpentine and talc, and often has an olive-enclosed texture. Pyroxene is mainly clinopyroxene, but often eroded into tremolite and actinolite, in the form of hypidiomorphic–euhedral columnar or granular, with a particle size of 5–10 mm and the long axis with directional distribution characteristics.

2.3.4.2 *Serpentinized dunite*

The serpentinized dunite and harzburgite are closely coexisted, and are produced in the form of lenticular blocks and slices. The rocks are dark grayish-black and grayish-green, with xenomorphic–granular texture and massive structure, strong serpentinization, directional arrangement of minerals, and development of mylonitic foliation. Bean-like and pod-like chromite are common (Figure 2-14) in the rock. The main minerals are olivine (30%–40%), serpentine (50%–60%), orthopyroxene (2%–3%), and chromite (1%–3%). Olivines are in xenomorphic-granular, with coarse crystal size up to 3–5 mm. Along the netted fractures, most olivines are altered into serpentine and talc in the shape of residual islands, or partly arranged linearly. Orthopyroxenes show hypidiomorphic–xenomorphic granular textures, with a particle size of 1–3 mm, completely metasomatized by serpentine, tremolite, and chlorite, with orthopyroxene pseudomorph preserved. It is occasionally seen that some columnar pyroxenes were replaced by serpentine into sericite, and tremolite is seen locally interspersing and enclosing olivine. With the enhancement of metasomatism and metamorphism, olivine shows transformation into tremolite, serpentine, clinohumite, magnesite, and especially talc, with the rock color changes from dark green to grayish-black and grayish-green.

2.3.4.3 *Metadiorite*

It is mainly distributed in the south of the serpentinized dunite and harzburgite, and occurs as intrusion or dyke. The rock is dark gray, with variable palimpsest cumulate texture, layered rhythmic structure, and massive structure. Some of rocks have undergone strong ductile deformation and show typical strip-eyeball structure. Palimpsest-gabbro

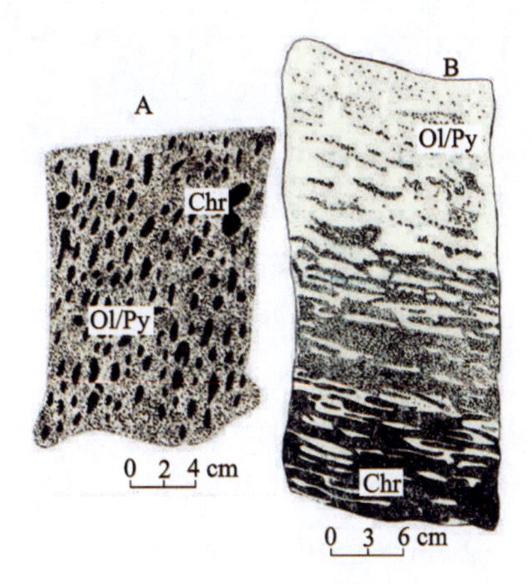

Chr. chromite; Ol/Py. olivine/pyroxene.

Figure 2-14 Chromite structure at Taipingxi, Hubei

(according to the internal data of Hubei Institute of Geological Science, 1973)

A. pisolitic chromite; B. massive–disseminated chromite

texture can be seen under the microscope. The main minerals are magnesium ordinary hornblende (40%–45%), basic plagioclase (40%–45%), augite (3%–5%), and magnetite (1%–2%). Augite is generally euhedral columnar or tabular, with a grain size of 5–8 mm. Most of the pyroxenes retrograded into hornblende, fibroamphibole, epidote, chlorite, etc. A few augite grains occur as island-like residues, often embedded with euhedral columnar plagioclase, and some are semi-embedded or resorption embayed textures. Plagioclase is mainly labradorite, in highly idiomorphic columnar shape, and with a particle size generally of 3–5 mm, which is slightly smaller than that of the pyroxene. Hornblende is mainly retrograded from pyroxene and in the shape of hypidiomorphic columnar or granular with a particle size of 2–3 mm.

2.3.4.4 *Metadiabase*

It is mainly distributed on the south side of the serpentinized dunite and harzburgite, closely coexisted with the metagabbro, intersecting and cutting each other, and occurs as veins and dykes. The rock is dark grayish-green, with palimpsest-gabbro or palimpsest-diabase texture and massive structure. Some of the rocks exhibit banded structure due to strong ductile deformation. The main minerals are augite (35%–40%), basic plagioclase (40%–45%), ordinary amphibole (5%–10%), and magnetite (1%–2%). Augite is generally heteromorphic and irregular, with a particle size of 1–2 mm. Most of them retrograded into hornblende, epidote, chlorite, etc., and a few remain in the shape of isolated islands. Plagioclase is mainly labradorite, column–granular in shape, euhedral–hypidiomorphic, with a particle size generally of 0.5–1 mm.

2.3.4.5 *Metabasalt*

It is mainly distributed on the northern side of the serpentinized dunite, harzburgite, metagabbro, and metadiabase, and occurs in a layered manner. The rock is dark gray, with palimpsest-porphyritic texture, stripe-banded structure, and has generally experienced ductile deformation and metamorphism. The grain size of the plagioclase porphyroblast is generally 2–4 mm. Some of the porphyroblasts are amphibole phenocrysts or amphibole aggregates, but still preserve the morphological characteristics of pyroxene. The matrix is composed of actinolite, labradorite or bytownite and sericite, with the particle size generally of 0.1–0.3 mm. The main mineral contents are magnesium ordinary hornblende (40%–45%), basic plagioclase (35%–40%), diopside (1%–2%), quartz (5%–10%), sericite (2%–3%), and magnetite (2%–3%). Magnesium ordinary amphibole is in short column shape, with mostly smooth grain edge, and develops with obvious wavy extinction. Occasionally, diopside metasomatized residual crystals that retain the shape of short columnar pyroxene can be observed. Plagioclase is plate-shaped, mostly metasomatized by albite or oligoclase, sericite and chlorite, but preserves the pseudomorph of plagioclase. Quartz is often lens-shaped or lentil-shaped, with directional arrangement obvious wavy extinction, and sub-grains developed. The albite or oligoclase is in the form of granular and lenticular aggregates, arranged in a directional arrangement, indicating that the metabasalt has undergone strong ductile shear deformation.

2.4 Metamorphic Rocks and Metamorphism

The metamorphic rocks in the Huangling Dome of the Three Gorges of the Yangtze River are mainly regional metamorphic rocks exposed in the Precambrian crystalline basement, followed by contact-thermal and dynamic metamorphic rocks.

2.4.1 Paleoproterozoic regional metamorphic rocks

The Paleoproterozoic high-grade metamorphic rocks are mainly distributed in the Yemadong Rock Formation, the Huanglianghe Rock Formation, the Li'erping Rock Formation, the Xiaoyucun Formation, and the Paleoproterozoic mafic–ultramafic rocks and granites in the core of the Huangling Dome. According to the deformation and metamorphism conditions, and the differences in mineral texture and structure, the common regional high-grade metamorphic rocks in this region can be divided into eight categories (Table 2-4).

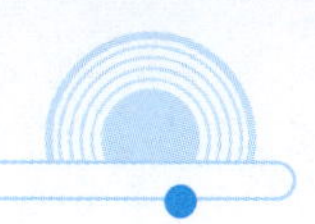

Table 2-4 Main types of metamorphic rocks in the northern part of the Huangling Dome

Rock Classification		Common Lithology	Protolith Characteristics
Schist	Mica (quartz) schist	dolomite quartz schist, garnet two-mica schist	muddy sandstone, quartz sandstone or mixed sandstone
	Al-rich schist	two-mica schist, graphite-bearing two-mica schist	clay siltstone, organic claystone
	Graphite schist	graphiteschist, graphite-bearing two-mica schist	high-organic mudstone
	Greenschist	chloritebiotite schist, zoisite-containing tremolite schist, epidote amphibolite schist	porphyric basalt
Granulite		biotite granulite, hornblende-plagioclase granulite, garnet-bearing plagioclase granulite	feldspathic sandstone, quartz sandstone, dacite volcanic rock
Gneiss	Al-rich gneiss	graphite, garnet, sillimanite-bearing biotite plagiogneiss, graphite, garnet-bearing biotite plagiogneiss	clay silt stone, organic claystone
	Plagiogneiss	garnet-beraing biotite plagiogneiss, hornblende plagiogneiss	dacite tuff
	Granitic gneiss	tonalitegneiss, trondhjemite gneiss, granodiorite gneiss, monzogranite gneiss	tonalite, trondhjemite, granodiorite, monzogranite
Amphibolite		quartz plagioclzse amphibolite, garnet amphibolite, biotite plagioclase amphibolite	basic volcanic rock, dolerite, calcareous sedimentary rock
Quartzite		hornblende quartzite, garnet quartzite, feldspar quartzite	quartz sandstone
Marble, Calcium-Silicate Rocks		tremolite marble, olivine marble, tremolite, diopside scapolitite, tremolite bistagite	dolomite limestone, marlstone, calcareous siltstone
Meta-Mafic–Ultramafic Rocks		(talcificated) serpentinite, diopside, chlorite tremolite schist	gabbro, pyroxenite, dolerite, augite peridotite, peridotite
Granulite	Mafic granulite	hypersthene granulite, hypersthene plagioclase hornblende, hypersthene-bearing garnet-hornblende plagiogneiss	basicrocks (dykes or interbeds)
	Argillaceous granulite	corundum-bearing sillimanite schist, eclogite	kaolinite claystone

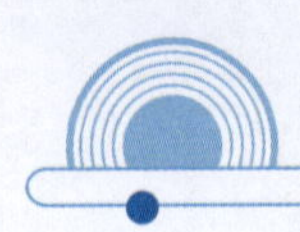

2.4.1.1 *Schist*

The schists in the core of the Huangling Dome are relatively well-developed, and mainly distributed in the Huanglianghe Rock Formation. They can be divided into four categories according to their mineral assemblages.

1. Mica (quartz) schist

It is rarely distributed in the Huanglianghe Rock Formation. Common lithologies are two-mica schist and garnet-bearing two-mica quartz schist. The rocks are characterized by directional arrangement of mica and squashed quartz. They consist mainly of mica, quartz, and minor sillimanite or garnet. The accessory minerals are mainly zircon, apatite, pyrite.

Sillimanite-bearing two-mica quartz schist: It shows an epidoblastic and lamellar texture. The main minerals are quartz (65%), biotite (7%), muscovite (25%), and sillimanite (3%). The biotite and muscovite are arranged in a directional arrangement to form flakes. Some biotite grains were replaced by muscovite through Fe precipitation. Quartz is distributed among flake minerals in the form of crystals. Sillimanite is hair-like, needle-like, and bundle-like, and often wrapped by quartz. The overall arrangement of sillimanite is uneven. The protolith of this type of schist is quartz sandstone.

2. Al-rich schist

It is mainly distributed in the Huanglianghe Rock Formation. The common lithologies are graphite-bearing staurolite (or sillimanite, andalusite) two-mica schist and graphite-bearing staurolite (or sillimanite, andalusite) two-mica quartz schist. The rock is light gray to dark gray, with a lepidoblastic, porphyroblastic, and lamellar texture. The main minerals are biotite, muscovite, and quartz with a varying amount of graphites, Al-rich minerals (andalusite, staurolite, sillimanite), and plagioclases. K-feldspar is absent or very minor. Protolith of Al-rich schist is clayed siltstone or mudstone.

Graphite-bearing andalusite staurolite two-mica schist: It is mainly distributed in the Huanglianghe Rock Formation. The rock is dark gray–grey and fine-grained, with a lepidoblastic and lamellar texture and occasionally a banded structure. The main minerals are quartz (20%–45%), biotite (20%–45%), muscovite (20%–30%), plagioclase (5%–20%), andalusite (0.5%–8%), staurolite (0.1%–8%), and sillimanite (0.1%–1%). A few rocks do not contain sillimanite or plagioclase.

Graphite-bearing sillimanite two-mica quartz schist: It is mainly distributed in the Huanglianghe Rock Formation. The rock is light gray, with a porphyroblastic, lepidoblastic, and lamellar-banded texture. The main minerals are quartz (50%–55%), muscovite (5%–45%), biotite (5%–20%), graphite (1%–5%), sillimanite (10%–20%), and garnet (1%–3%).

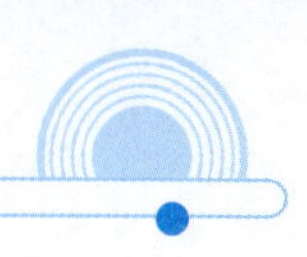

The plagioclase or quartz are often aggregated into bands or lenses, and parallel to dark bands. Due to the migmatization, vermicular structures are developed on the edges of plagioclase or quartz.

3. Graphite schist

It is mainly distributed in the Huanglianghe Rock Formation. Common lithologies are graphite schist and graphite two-mica schist, which are layered or lenticular, accompanied by Al-rich schist and marble. The rock is black, with a lepidoblastic and lamellar texture. The main minerals are biotite, muscovite, and graphite, with a small amount of feldspars, quartzs, and garnets, among which graphite is 20%–40%. For areas (such as Sanchaya and Houshan Temple) where graphite is as high as 60%, graphite deposits are formed. It has been identified that graphite contains micro-paleo fossils (Yichang Geological Survey Team, 1987), indicating its organic origin. Thus, the protolith of the graphite schist is organic mudstone.

Graphite two-mica schist: It is layered or lenticular is coexisted with with Al-rich schist, gray-bright to gray, with a lepidoblastic and lamellar texture. The main minerals are graphite (20%–45%), muscovite (5%–25%), biotite (10%–40%), and quartz (2%–20%). Mica and graphite intersect to form rock flakes. Graphite wafers are large, and quartz crystals are fine-grained. Pyrite is scattered in the rock, mostly limonite mineralized.

4. Greenschist

It is mainly distributed in the Yemadong Rock Formation and Li'erping Rock Formation. The common lithologies are epidote amphibole schist, chlorite amphibole biotite schist, and zoisite-bearing tremolite schist. The main minerals are actinolite-tremolite, chlorite, amphibole, biotite, and plagioclase. The protolith of the greenschist is porphyric basalt.

Epidote amphibole schist: The rock has a granular columnar lepidoblastic texture and a lamellar texture. The main minerals are amphibole (50%–70%), plagioclase (10%–20%), and epidote (12%–25%), with a small amount of biotites, calcites, ilmenites, and sphenes. Hornblendes are light green in color and are arranged uniformly. Biotite inclusions can be seen inside, and colorless hornblende coronas are occasionally occurred on the edges of the hornblendes. Epidote is fine-grained and distributed closely with the hornblende. The plagioclase is mainly of albite ($An<8$) and distributed in the interstices of hornblende. Ilmenite or sphene are scattered together, and there are abundant albite and amphibole inclusions inside. The protolith of the epidote-amphibole schist could be mafic rock.

Chlorite amphibole biotite schist: The rock has a granular columnar crystalloblastic texture and a lamellar structure. The main minerals are amphibole (5%–10%), plagioclase (10%–20%), biotite (20%–45%), chlorite (1%–8%), and quartz (1%–5%). The

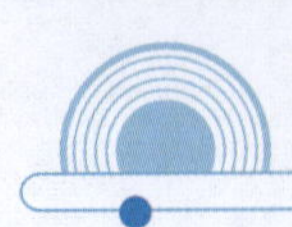

amphibole is light green and arranged uniformly. The biotite is brown and distributed together with amphibole. Plagioclase is scattered in the interstices of amphibole or biotite grains, and experienced extensive serpentinization or albitization. Light red biotites or yellowish-green chlorites are abundantly seen in the edges of pale green amphiboles or brown biotites due to the retrograde metamorphism. The protolith of the chlorite amphibole biotite schist is mafic rocks.

Zoisite-containing tremolite schist: It is widely distributed in the Yemadong Rock Formation, the main minerals are zoisite-tremolite (57%–70%), plagioclase (20%–40%), zoisite (10%–20%), and chlorite (1%–5%), with a small amount of biotites and ilmenites. The rock has a granular columnar crystalloblastic texture and a lamellar structure. Zoisite is distributed in agglomerated form in the gap between actinolite and tremolite. Plagioclase is strongly sericitized or albitized, but preserves slaty or columnar crystals. The protolith is mafic rocks.

2.4.1.2 *Leptynite*

It is distributed in the Huanglianghe Rock Formation and the Yemadong Rock Formation. In the former Huanglianghe Rock Formation, the common lithology is biotite plagioclase leptynite; while in the latter, the lithology is generally hornblende plagioclase leptynite. They all have a fine-grained and uniform-grained poikiloblastic texture, and a massive and gneissic structure. The main minerals are amphibole, biotite, plagioclase, and quartz, with a small amount of diopsides and almandines. The accessory minerals are zircon, apatite, ilmenite, hematite, and limonite. Zircon is generally rounded. The protolith is feldspathic sandstone and intermediate igneous rock.

1. Biotite plagioclase leptynite

It is distributed in the Huanglianghe Rock Formation, exhibits thick layered feature and coexisted with Al-rich gneiss. The main minerals are plagioclase (44%–60%), quartz (20%–30%), and biotite (15%–25%), with variable garnet and graphite. Plagioclase is mostly sericitized or epidotized, and the garnet and biotite are mostly replaced by chlorite. The protolith is feldspathic sandstone.

2. Hornblende plagioclase leptynite

It is distributed in the Yemadong Rock Formation, sandwiched in plagioclase gneiss or interbedded with plagioclase amphibolite. The main minerals are plagioclase (30%–50%), quartz (15%–40%), amphibole (10%–15%), and biotite (5%–10%), with variable actinolite-transparent tremolite and epidote. Hornblende is blueish-green, and the edges are often replaced by tremolite or biotite. Plagioclase is mostly sericitized or epidotized. The protolith is dacite tuff.

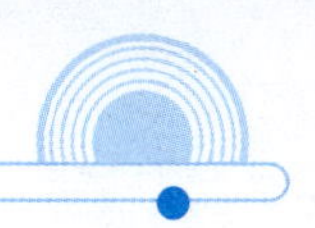

2.4.1.3 *Gneiss*

In the area, gneiss is relatively well developed and can be divided into two categories: paragneisse and orthogneisse. Paragneisses is mainly distributed in the Huanglianghe Rock Formation and the Yemadong Rock Formation; orthogneisses is distributed in the Dongchonghe gneiss complex, the Bashansi gneiss complex, and the Saijiachong gneiss. According to different mineral assemblages, gneiss can be divided into Al-rich gneiss, plagiogneiss, and granitic gneiss.

1. Al-rich gneiss

It is distributed in the Huanglianghe Rock Formation, with a fine-grained lepidoblastic texture and a gneissic structure. In this type of rocks, the brownish-red biotite (relatively high content of 20%–30%), garnet (sometimes as high as 10%–20%), and fine needle columnar sillimanite (sometimes more than 10%–15%) porphyroblasts are commonly observed. The felsic minerals are oligoclase, and a varying amount of quartzs and K-feldspars. The content of graphitic flakes is 1%–3%. Detailed speaking, the most common types of the Al-rich gneiss are graphite-bearing sillimanite garnet biotite plagiogneiss and graphite-bearing biotite plagiogneiss. There are mica schists interbedded in the above-mentioned gneissic rocks. The mica schists are generally in dark colour, with fine-grained lepidoblastic texture. The biotite and muscovite together in the mica schists have a content of 40%–50%, and the rest are mainly of quartz and a small amount of acidic plagioclases. Some of the mica schists contain porphyroblasts such as andalusite, garnet, and staurolite. The most commonly seen mica schists are graphite-bearing andalusite garnet two-mica schist and andalusite staurolite two-mica schist. In addition, two-mica schist is high in graphite content, and (biotite) graphite schist as graphite ore are also commonly developed. They have a very fine-grained (0.02–0.03 mm) lepidoblastic texture and a near phyllitic structure. The protolith of the mica schists is argillaceous siltstone-mudstone with organics.

Graphite-bearing biotite plagiogneiss: The main minerals are quartz (10%–27%), plagioclase (25%–55%), biotite (7%–15%), and graphite (3%–7%). The quartz is often squashed with unequal grains, broken and fine-grained inside, and displays blastomylonitic texture due to subsequent recrystallization. The plagioclase is fine-grained and exhibits obvious sericitization. The biotite shows pleochroism of marron and pale-yellow, accompanied by exsolution of iron. The graphite is banded and distributed along with the biotites.

Graphite-bearing garnet sillimanite biotite plagiogneiss: The main minerals are quartz (1%–23%), plagioclase (25%–50%), biotite (7%–25%), sillimanite (2%–21%), and garnet (5%–20%). Sillimanite has two forms: hair-like and prismatic. The former form

often occurred as reaction rim of the biotite, while the latter form is in equilibrium contacted with biotite.

Graphite-bearing two-mica plagiogneiss: The rock has a granular lepidoblastic texture and a gneissic-banded structure. The main minerals are quartz (30%–57%), plagioclase (30%–36%), biotite (15%–25%), muscovite (10%–30%), and graphite (<5%). Biotite often unequally transformed into muscovite due to retrograde metamorphism.

Garnet-bearing andalusite staurolite biotite plagiogneiss: The rock has a granular lepidoblastic or porphyroblastic texture and a gneissic-banded structure. The main minerals are quartz (30%), plagioclase (28%), biotite (25%), muscovite (5%), and alusite (5%), staurolite (3%–5%), and garnet(<5%), with scattered graphite, zircon, apatite, pyrite, tourmaline, etc. The felsic minerals are mostly aggregated in bands or lenses and distributed along the gneissic schistosity, and contain many mineral inclusions including mica and graphite.

2. Plagiogneiss

It is distributed in the Huanglianghe Rock Formation and the Yemadong Rock Formation. In the Huanglianghe Rock Formation, the common rock type is garnet-bearing biotite plagiogneiss, which is mainly composed of biotite, felsic minerals, and a small amount of garnets. Overall, the garnet-bearing biotite plagiogneiss has mainly a granular lepidoblastic texture, with some parts preserve well-developed blastopsammitic texture. The protolith of the garnet-bearing biotite plagiogneiss is probably feldspathic quartz sandstone. The common rock type of the Yemadong Rock Formation is hornblende plagiogneiss and biotite hornblende plagiogneiss. According to the geochemical composition, their protolith is probably of dacite volcanic tuff.

Garnet-bearing biotite plagiogneiss: It is distributed in the Huanglianghe Rock Formation. The main minerals are quartz (30%–35%), plagioclase (30%–45%), garnet (10%–15%), and biotite (5%–20%), with a small amount of epidotes and ilmenites. Garnets are oblate and lenticular-shaped, and metasomatized by chlorite along the cracks to form a reticulate structure. Plagioclase is strongly sericitized, leaving only residual granular crystals, while biotite is strongly chloritized, leaving only red residual crystals in the core. The protolith of the garnet-bearing biotite plagiogneiss is probably feldspathic quartz sandstone.

Hornblende plagiogneiss: It is distributed in the Yemadong Rock Formation, often interbedded with amphibolite. The main minerals are plagioclase (30%–56%), quartz (15%–30%), hornblende (3%–36%), and biotite (5%–10%). The minerals coexist in equilibrium. Worm-like quartz texture is developed at the margin of the hornblende plagiogneiss due to anatexis. The protolith of the hornblende plagiogneiss is probably dacitic volcanic tuff.

3. Granitic gneiss

It is a tonalitic-trondhjemitic-monzogranitic gneiss that intruded into the surrounding rock. The rock has the appearance of granitic pluton on the field outcrop, and is developed with a large number of mafic enclaves that represent the residues of partial melting. Detailed descriptions for this dranitic gneiss are listed in Section 2.3.

2.4.1.4 *Amphibolite*

Amphibolite is distributed in the Huanglianghe Rock Formation, the Li'erping Rock Formation and the Yemadong Rock Formation. There are three types of commonly developed rocks.

1. Quartz plagioclase amphibolite

It is developed in thin beds or interlayers, interbedded between the Al-rich gneiss of the Huanglianghe Rock Formation and the Yemadong Rock Formation, and is scattered in the region. The rock has a coarse-grained and columnar-blastic texture and a speckled or schistose structure. The main minerals are quartz (10%–20%), hornblende (30%–45%), and plagioclase (20%–40%), with a varying amount of garnets, biotites, and diopsides. The hornblende is columnar and light green in color, with pleochroism of light green to light yellowish-green. Plagioclase is granular, and its surface is obviously sericitized and chloriteized. Quartz is inequigranular, and distributed in the gap between the plagioclase and hornblende.

2. Garnet amphibolite

It is distributed in the Yemadong Rock Formation and Li'erping Rock Formation. The garnet amphibolite in the Yemadong Rock Formation occur as variable sizes of xenoliths within the Dongchonghe gneiss complex and is mostly interbedded with gneiss and fels. And the garnet amphibolite in the Li'erping Rock Formation is developed with single lithology, but variable thicknesses. The garnet amphibolite is mainly composed of hornblende (45%–60%), garnet (5%–15%), and plagioclase (15%–40%), and has a fine–medium-grained crystalloblastic texture and a banded or speckled structure with strong anatexis. The hornblende in the garnet amphibolite is mainly blueish-green. The protolith of the garnet amphibolite is probably mafic volcanic tuff.

3. Biotite plagioclase amphibolite

It is occurred as dyke and constitutes a part of the Hetaoyuan mafic–ultrabasic rocks. The rock exhibits a blastodiabasic texture with relic clinopyroxene can be observed, and a

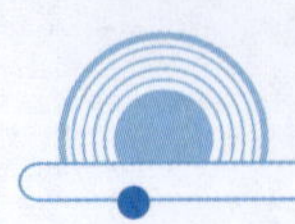

massive structure with its edge been schistositized. The mineral grain size of the rock is coarse in the center but fine at the edge. Accessory minerals are sphene, magnetite, ilmenite, and apatite. The geochemical composition shows that the amphibolites are mafic, and thus the protolith of this biotite plagioclase amphibolite is probably diabase and gabbro-diabase.

2.4.1.5 *Marble and calcium-silicate rocks*

They are mainly distributed in the Huanglianghe Rock Formation and occur lenticularly or interlayerly.

1. Marble

It mainly consists of tremolite marble and olivine marble. Their main minerals are dolomite, tremolite, and calcite, with a varying amount of diopsides, grossularites, scapolites, olivines, and graphite flakes. The accessory minerals are a few and mainly of zircon, magnetite, and cordite. These marbles are often coexisted with the quartzite, graphite schist, and Al-rich rocks in the field. The protolith of these marbles should be argillaceous dolomite limestone.

Tremolite marble: The rock is white with a fine-grained granoblastic texture and a massive structure. The main minerals are dolomite (30%), calcite (60%), and tremolite (10%). All of the minerals coexisted in equilibrium.

Olivine marble: The rock is pale yellow to off-white, with a fine-grained granoblastic texture and a massive structure. The main minerals are dolomite (40%–60%), calcite (10%–20%), and olivine (10%–25%), with a small amount of phlogopites, grossularites, and diopsides. Olivines coexisted in equilibrium with calcites and dolomites, and are commonly serpentinized.

2. Calcium-silicate rocks

This type of rock commonly consists of diopside scapolitic rock, tremolitic rock, tremolite diopsidite, and plagioclase diopsidite. The rock is off-white, with a fine-grained granoblastic texture and a massive structure. Their main minerals are calcium-magnesium-silicate minerals, such as diopside, sclerite, tremolite, and zoisite, with a variable amount of plagioclases, quartzs, and graphites. The protolith of this type of rocks is dolomitic limestone and calcareous siltstone.

Diopside scapolitic fels: It is grayish-white, with a coarse-grained granoblastic texture and a massive structure. The main minerals are diopside (38%–42%) and sclerite (50%–55%), with a small amount of plagioclases and accessory minerals.

Tremolitic rock: It is pale green, with a coarse-grained granoblastic texture and a

massive structure. The main minerals are tremolite (85%) and dolomite (10%), with a small amount of olivines and plagioclases.

Tremolite-diopsidite: It is dark green, with a coarse-grained granoblastic texture and a massive structure. The main minerals are tremolite (5%–48%) and diopside (50%–69%), with a small amount of quartzs and micas.

2.4.1.6 *Quartzite*

The quartzite occurs mainly as lenticular form in the Al-rich rocks and graphite-bearing schist (gneiss) of the Huanglianghe Rock Formation. The most commonly seen rocks include amphibolite quartzite and garnet-bearing quartzite, as well as feldspar quartzite, all of which have uniform massive structure, with quartz content of more than 80% and also contain a certain amount of garnets, plagioclases, graphites, and hornblendes. The accessory minerals are zircon, apatite, and magnetite. The protolith of the quartzite is quartz sandstone.

2.4.1.7 *Meta mafic–ultramafic rocks*

They are mainly distributed in the ultramafic rocks and can also be found in the Yemadong Rock Formation. The common lithologies are chlorite tremolite schist, diopsidite, talcose rock, serpentinite, serpentinized peridotite, and pyroxene peridotite. Their protoliths are gabbro, peridotite, and pyroxene peridotite.

2.4.1.8 *Granulite*

The granulite can be mainly divided into pelitic granulite (garnet-sillimanite-quartzite) and mafic granulite.

1. Pelitic granulite

In this study, rocks that contain a large amount of sillimanites, garnets, and quartzs (also called garnet sillimanite quartzite in a broad sense), and similar Al-rich rocks are collectively referred to as garnet sillimanite quartzite. This kind of rocks generally have Al_2O_3 content of 22.2%–29.2% and are classified as typical khondalite series. They are distributed in the Huanglianghe Rock Formation, and the common rock types are corundum-bearing garnet sillimanite schist or gneiss, sillimanite-bearing staurolite andalusite kyanite garnet schist, and garnet sillimanite quartzite. They are interbedded or developed in lenticular form in Al-rich schist or gneiss, and often coexist with quartzite. They have a gneissic-massive-taxitic structure. Their protolith may be aluminum-siliceous component cemented kaolinite claystone.

1) Corundum-bearing garnet sillimanite schist or gneiss

It is grayish-white, with a fibrousblastic–porphyroblastic texture and a schistose or gneissic

structure. The main minerals are sillimanite (20%–40%), plagioclase (25%–45%), garnet (5%–10%), biotite (2%–3%), and corundum (5%–10%), with a small amount of accessory minerals such as zircon and rutile. Thc sillimanite occurs as prism-shaped aggregation and distributes in consistent along with the schistosity of the parent rock.

2) Sillimanite-bearing staurolite andalusite kyanite garnet schist

It is grayish-white, with a porphyroblastic texture, and a schistose structure. The main minerals are sillimanite (8%), staurolite (5%), andalusite (5%), kyanite (11%), garnet (45%), quartz (14%), muscovite (8%), and biotite (2%), with a small amount of accessory minerals such as zircon, ilmenite, and magnetite.

3) Garnet silimanite quartzite (narrow sense)

It is light brown, with a fibrousblastic–porphyroblastic texture and a taxitic-massive structure. The main minerals are quartz (10%–40%), sillimanite (10%–38%), staurolite (0–5%), andalusite (0–5%), garnet (25%–60%), feldspar (0–2%), and biotite (2%–3%). The rock contains extremely low content of accessory minerals.

2. Mafic granulite

It is mainly distributed in Qinjiaping–Zhoujiahe–Tandanghe, and also occurred in Erlangmiao and Lijiawuchang. It is usually lenticular in shape and interbedded within the amphibolite facies metamorphic rock of the Huanglianghe Rock Formation. The following lithologies are commonly seen.

1) Hypersthene-bearing plagioclase amphibolite

It is dark gray, with a medium–fine-grained granoblastic texture and a banded-speckled structure. The main minerals are amphibole (40%–62%), hypersthene (2%–7%), plagioclase (25%–44%), quartz (3%–4%), and garnet (1%–4%). Hypersthene is granular, from light green to light red in color. A small amount of fine-grained relic amphibole inclusions are occasionally found in the hypersthene.

2) Hypersthene granulite

It is dark brown, with a coarse-grained granoblastic texture and a taxtic structure. The main minerals are hypersthene (36%–60%), garnet (1%–34%), and quartz (<1%). The garnets are mostly distributed in intermittent bands to form a granulitic texture. Hypersthene is metasomatized by tremolite but maintains the metasomatic pseudomorph texture.

3) Hypersthene biotite plagiogneiss

It is grayish-white, with a fine-grained porphyroblastic texture and a banded structure. The main minerals are biotite (<15%), plagioclase (55%), quartz (5%), hypersthene (5%–10%), and garnet (15%). Amphibole reaction rim is commonly found around the hypersthene. The rock is characterized by the presence of hypersthene and garnet, with or

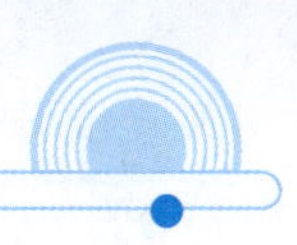

without quartz. The protolith of the hypersthene biotite plagiogneiss could be calcium silicate rocks and mafic rocks.

Overall speaking, the mafic granulite usually retrograded into amphibolite, with some parts gradually transformed into biotite plagiogneiss. The protolith of the mafic granulite should be mafic rock.

2.4.2 Contact metamorphic rocks

The contact metamorphic rocks in the Huangling Dome area are mainly skarn formed by contact metamorphism, which are found in the Songshuping and Liujiawan areas. Skarn-type copper-molybdenum mineralization can be seen in the contact zone between the Xiping unit of the Huangling granite rock foundation and the marble in the second member of the Xiaoyucun Formation.

2.4.2.1 *Diopside skarn*

It is mainly composed of diopside (over 90%), quartz (0–5%), actinolite, and calcite, with trace molybdenite, chalcopyrite, magnetite, and other metallic minerals. Diopside is mostly replaced by amphibole and chlorite.

2.4.2.2 *Diopside quartz skarn*

It is mainly composed of quartz (75%–80%), diopside (about 15%), and plagioclase (about 5%), and a small amount of garnets, chlorites, biotites, etc. Quartz is granular, and diopside is commonly occur in granular or as columnar aggregates.

2.4.2.3 *Garnet epidote diopside skarn*

It is mainly composed of diopside and epidote (70%–75%), garnet (about 15%), pyrite, and sphene. Most of the diopsides and epidotes are xenomorphic and granular, and a few are short-prismatic in shape. The garnets are unevenly occurred as granular aggregates.

2.4.2.4 *Quartz epidote diopside skarn*

It is mainly composed of diopside and epidote (about 50%), quartz (40%–45%), actinolite, sphene, apatite, pyrite, etc. Diopside and epidote are developed in hypidiomorphic-short prismatic grains and xenomorphic-granular crystals. Quartz is evenly occurred in xenomorphic-granular aggregates.

2.4.3 Dynamic metamorphic rocks

During the secular evolution, the Huangling Dome area has experienced multistage ductile, brittle-ductile deformation, and metamorphism, and brittle fracture and modification,

forming a variety of dynamic metamorphic rocks. They are generally linear or zonal distributed with greatly variable width and distribution length. According to the mechanism and tectonic setting of the dynamic metamorphism, the dynamic metamorphic rocks can be subdivided into three types, including ductile metamorphic rocks, brittle-ductile metamorphic rocks, and brittle metamorphic rocks (Table 2-5).

Table 2-5 Classification of dynamic metamorphic rocks

Types	Rocks	Main Features	Tectonic Setting
Ductile metamorphic rocks	mylonitized rocks, protomylonites, mylonites, and ultramylonites	typically mylonitic texture, fluxion structure, various plastic kinematic signs developed, accompanied by amphibole stretching lineation and biotite stripe lineation	greenschist facies–amphibolite facies
	tectonic schist	strong straight gneissic structure, banded structure, mainly of blastomylonitic texture, with rectangular felsic minerals occurred in stripped shape, accompanied by amphibole stretching lineation	amphibolite facies
Brittle-ductile metamorphic rocks	greisenic tectonic schist, felsic tectonic schist	macroscopically foliaceous and tile structure, accompanied by felsic stretching lineation and sericite stripe lineation. Micas in the rock usually occur as fishes in shape due to plastic deformation, shear deformation, and brittle deformation under pressure-soluble fracture	low greenschist facies
Brittle metamorphic rocks	fault breccia, cataclastic rocks, porphyroclastic rocks, clastic felses, and crushed rocks	cataclastic texture, massive structure. Due to multi-stage modification, "gravel-enclosed gravel" texture, accompanied by scratch lineation and traction folds, are usually observed	hydrothermal alteration

2.4.3.1 *Ductile metamorphic rocks*

Ductile metamorphic rocks are widely developed in the high-grade metamorphic rock (granitic gneiss) in the northern part of the Huangling Dome, and in the ophiolitic mélange in Miaowan of the southern part of the Huangling Dome. The main rock types are tectonic gneiss, mylonite, and blastomylonite.

1. Tectonic gneiss

The tectonic gneiss, which is characterized by strong straight gneissose structure in the field outcrops, is mainly composed of felsic tectonic gneiss. In the felsic gneiss, the felsic minerals are generally strongly compressed, elongated, and recrystallized to form rectangular ribbons, needle–columnar amphiboles, and biotite scales which distributed along the gneissic orientation. Deformed felsic veins are seen in asymmetrical fold, showing signs of kinematics. These gneiss are mostly formed under amphibolite-granulite facies conditions.

2. Mylonite

The mylonite shows a typical mylonitic texture and a fluxion structure in both the field outcrops and the microscopic scale. The common rocks are felsic mylonites and amphibolic mylonites. They can be further divided into mylonitized rocks, protomylonites, mylonites, and ultra-mylonites according to the content of mylonitic phenocrysts. The mylonitic porphyroclastic residuals are mainly composed of K-feldspar, plagioclase, quartz, and amphibole. A variety of micro-kinematic signs are developed in the mylonites, such as S-C fabric, "σ"-type rotating phenocrysts, asymmetric pressure shadows, and feldspar or mica bookshelf structure. The amphibole stretching lineation and biotite stripped lineation are commonly developed in the field outcrops, indicating a formation condition of low amphibolite facies and high greenschist facies.

1) Mylonitized rock

It is the rock that has a mylonitic texture but with the lowest degree of deformation in the category of mylonites. Usually in this rock, the felsic minerals are rimmed by sub-granulation. while the quartzes dispaly a flow texture and show undulatory extinction. In those residual porphyroclasts, the plagioclase are well developed with mechanical twins, and the cleavage lines of the biotites are slightly twisted. Common rock types include mylonitized monzogranites and mylonitized felses.

2) Protomylonites

These rocks have matrix content of 5%–10%, with obvious fluxion structure. The quartzes show undulatory extinction, and the plagioclase twins are bent. The cleavage lines of the biotite are bent and deformed, and occasionally core-mantle structure is developed in the boitites. The common rocks are felsic protomylonites and mafic protomylonites.

3) Mylonite

The mylonites are mostly outcropped in the NE or NWW shear zone in the studied area. The content of the matrix is more than 50%, while the porphyroclasts are mainly of plagioclase and K-feldspar. The residual porphyroclastic feldspars are augen and lenticular in shape. The quartzes are elongated with ribbon-like and undulatory extinction, and are

almostly recrystallized. Some of the feldspars occur as "σ"- or "δ"-type rotational porphyroclasts, while some residual porphyroclastic feldspars are micro-cracked with obvious bookshelf texture. The biotites form as mica fish under stress, and in some parts retrograded into chlorites. Therefore, the shearing characteristics and histories of the shear zone can be investigated based on the development of "σ"-or "δ"-type-rotational porphyroclastic feldspar, the oblique texture in the feldspar, the mica fish, and the S-C fabrics in the mylonites. The common mylonites are granitic mylonite and felsic mylonite.

3. Blastomylonite

The blastomylonites are distributed within the metamorphic complexes in the Yemadong Cave area. They are the products of plastic deformation occurred in the middle deep depth. Most of the rocks have undergone substantial static recrystallization and are characterized by an obvious blastomylonitic texture, with felsic rectangular polycrystalline strips. According to the differences of the degree of mylonitization, the rocks can be subdivided into two types of blastoprotomylonite and blastomylonite.

1) Blastoprotomylonite

This kind of rock formed through plastic flow under stress followed by static recrystallization and contains abundant blastoporphyroclasts. The degree of plastic deformation differs significantly among different minerals. The quartzes are strongly plastically deformed, while the feldspar minerals are slightly plastically deformed in the lenticular or flattened granular form. The plagioclases are developed with mechanical twins. The main rock type is granitic blastoprotpmylonite.

2) Blastomylonite

The rock is characterized by strong mylonization, directional and banded structure. Mica fishes are commonly seen, and some of the biotites have undergone retrogradation. Residual porphyroclastic feldspars are occasionally seen as augenor lenticular shapes, with some grains occur as rotational crystals. The rock has experiencd substantial recrystallization, reflected by the widely occurrence of enlarged ductile matrix particles. The quartzes generally occur as polycrystalline bands, and are bent near the porphyroclasts, showing blastomylonitic texture. The particles of feldspars have been sub-granulated and are arranged in an orientation under shear stress. After static recrystallization, the feldspars display well observed triple point structure and distributed in a mosaic arrangement. In some rocks, the apatites is distrubuted in chains. The commonly outcropped rocks in the area are granitic blastomylonites.

2.4.3.2 *Brittle-ductile metamorphic rocks*

This type of rocks are mainly composed of tectonic schists, which are distributed along the

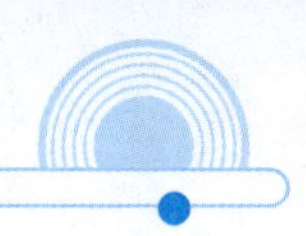

ductile shear zone. Some of the rocks are outcropped separately in a linear form, and some are superimposed on the early ductile shear zone. The most commonly developed rocks are two-mica tectonic schists, sericite quartz tectonic schists, and chlorite sericite tectonic schists.

A group of discontinuous cleavage, which is often inconsistent with the regional schistosity or gneissosity, can be observed on the filed outcrops, accompanied by stretching lineation of felsic minerals, streak lineation of biotite or muscovite. Under the microscope, there are "mica fish", pressure shadow, and shear pressure dissolution. These indicate a low-greenschist metamorphic condition.

1. Two-mica tectonic schist

It has a banded structure, with the bands consist by the mutual arrangement of micas and felsic minerals. The felsic minerals are strongly flattened and oriented, with undulatory extinction. The micas are lenticular or in the shape of fish, with bended cleavages. In addition, widespread cracks, which are filled by carbon particles, are found in the micas.

2. Chlorite sericite tetonic schist

It is strongly foliated, with extremely uneven interference colors in individual minerals. Most of the minerals are characterized by undulatory extinction. The micas and chlorites are elongated into ribbons, with their edges are irregular or jagged. The internal cleavages of the minerals are bended and are penetrated by quartz veins.

2.4.3.3 *Brittle tectonic rocks*

The brittle rocks are mostly distributed along late stage of brittle faults. They are generally of various forms of cataclastic rocks that mainly produced through the destruction and modification of the preexisted metamorphic rocks, sedimentary rocks, and granites during the uplift of the Huangling Dome since Middle Cenozoic. The rocks include fault breccias, cataclastic rocks, porphyroclastic rocks, clastic felses, and crushed rocks. Thermal alteration such as silicification and sericitization often occurred in these rocks. In regional large fault zones, it is common to see that the cataclastic rocks enclosing mylonites, or even the cataclastic rocks enclosing cataclastic rocks, such as the fault breccia enclosing porphyroclastic rocks and the porphyroclastic rocks enclosing clastic felses. These phenomena, which are also named as "gravel enclose gravel", indicating the multistage activities of the regional faults.

1. Fault breccias

These rocks have a gravel texture, with the breccias larger than 2 mm, and the content of

clastic-matrix is less than 30%. According to the shapes of the breccias, the fault breccias can be subdivided into extensional breccias and compressive breccias.

1) Extensional breccias

The breccia fragments are mostly angular, mixed in size and disorderly arranged. The cements are mainly of calcareous, argillaceous, ferruginous, and siliceous in composition. In addition, the autogenous fragments of breccias can also serve as cements.

2) Compressive breccias

The breccia fragments are lentil-shaped, from sub round to round. There is little disparity in the breccias. Relative to the extensional breccias, the compressive breccias have a higher amount of clastic-matrix but less cements. In addition, the clastic-matrix generally presents in directional arrangement.

2. Cataclastic rocks

They have a cataclastic texture. Usually, the porphyroclasts in the rocks have little displacement, and hence they can be roughly spliced. The cements filling between the porphyroclasts are manily of argillaceous, ferruginous, and siliceous in composition, and have a content of less than 50%.

3. Porphyroclastic rocks

They have a porphyroclastic texture, namely, the residual porphyroclasts are generally surrounded by broken particles and cushed materials that produced by fracture. The porphyroclasts are more than the clastic matrix. Most of the porphyroclasts have experienced displacement and rotation, but still preserve the primary nature and structure of the protolith to varying degrees. Edge granulation and tearing are very common in the porphyroclasts. Plastic deformation signs, such as deformation lamellaes and kink bands, can also be seen in the porphyroclasts.

4. Clastic felses

They have a clastic granular texture, with most of the minerals are broken into clastic grains or powder. The original texture and structure of the protolith are difficult to be identified. Both of the porphyroclasts and clastic grains have a low contents, and they also tend to be rounded, with plastic deformation can be observed.

2.5 Geological Structure

The formation and tectonic evolution of the South China Yangtze Craton's basement during the pre-Nanhuaian have been highly concerned by geologists domestic and abroad. However, since most of the Yangtze Craton is covered by sediments since the Nanhuaian, the composition, structure, and tectonic evolution characteristics of the pre-Nanhuaian basement are mainly based on deep geophysical data and the study of the rarely exposed pre-Nanhuaian basement (e.g., the Huangling Dome). It is generally believed that the Archean crystalline basement may be widespread in the Yangtze Craton, and it was formed by the accretion and collage of several micro ancient continental nuclei (花友仁, 1995; 袁学诚, 1995; 白瑾等, 1996; Zheng et al., 2006).

In recent years, three discoveries indicate that the basement of Yangtze Craton in pre-Nanhuaian not only has the geological record of Neoproterozoic subduction collision orogeny (Zhang et al., 2009; Qiu et al., 2011; Peng et al., 2012; Wei et al., 2012; Bader et al., 2013), but also has the important record of Paleoproterozoic subduction collision orogeny metamorphism and disintegration. These three discoveries are: the discovery and identification of the late Mesoproterozoic–the early Neoproterozoic (1.1–0.98 Ga) Miaowan ophiolite (彭松柏等, 2010; Peng et al., 2012) in the southern part of the Huangling anticline in the core of the Yangtze Craton; the discovery of the hidden Neoproterozoic or Paleoproterozoic subduction zone (董树文等, 2012; Dong et al., 2013) in the deep part of pre-Nanhuaian basement in Yangtze Craton; the confirmation of the Paleoproterozoic (2.0–1.95 Ga) high pressure granulite facies tectonic metamorphism and the 1.86–1.85 Ga pyrolytic magma events (凌文黎等, 2000; 熊庆等, 2008; 彭海文等, 2009; Peng et al., 2012; Yin et al., 2013) of the pre-Nanhuaian basement in the northern part of the Huangling anticline.

The regional tectonic evolution of the Huangling area in the Yangtze Craton can be roughly divided into two major tectonic evolution stages: the basement and caprock evolution. The pre-Nanhuaian basement has experienced two important subduction-collision orogenic collages. The late Neoproterozoic subduction collision orogenic collage finally formed the basic outline of the Yangtze Craton basement and entered the stage of stable sedimentary caprock tectonic evolution. In Late Mesozoic, the area began to be affected by the subduction of the Pacific Plate and the uplift of the Qinghai-Tibet Plateau, characterized by intracontinental compression-extension evolution. The main geological structural features at different stages of geological structural evolution are briefly described as follows.

2.5.1 Ductile deformation structure of the Archean granitic gneiss

The Archean granite gneiss (TTG) is mainly distributed in the northern part of the Huangling Dome. It is mainly characterized by ductile shear rheological deformation in the middle and deep layers, and has obviously different deformation and metamorphism characteristics from the overlying strata. Nearly east-west ductile flow and fold structures are developed, and rocks are generally subjected to regional metamorphism of amphibolite facies, accompanied by the formation of regional tectonic facies and gneiss. The main manifestations are as follows:

(1) In the Archean granitic gneiss (TTG gneiss), penetrating ductile shear zones and gneisses are widely developed. Felsic or granitic veins often form asymmetric shear folds with closed limbs (Figure 2-15). Due to the later structural deformation, the foliation is often deformed and displaced, and its restored attitude is 200°–240°∠30°–50°. Due to the strong shearing and stretching, some folds are broken, forming rootless folds with hinge zone significantly thickened. The amphibolite xenoliths or veins in the Dongchonghe gneiss complex often form structural lenses and boudinage structures with the axis of lens parallel to the gneiss foliation.

Figure 2-15 A field photo of the Archean granite gneiss in the southeast of Shuiyuesi
(photo by Han Qingsen)

(2) In the Archean granitic gneiss (TTG), ductile shear foliation and tectonic gneiss often develop felsic veins formed by partial melting. The veins often develop intermittently parallel

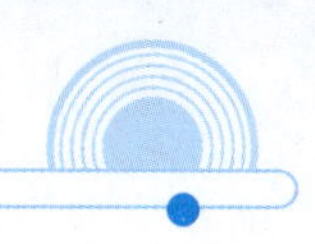

to the new foliation, and are darker. The stripes of mineral composition are arranged alternately to form a striped-banded structure, generally ranging from a few millimeters to a few centimeters in width. Veins often form rootless folds, boudinages, and structural lenses by ductile shear deformation. Gneissic is represented by the mineral reorientation of hornblende and biotite. The structural foliation in this period was deformed and replaced by subsequent structure, and only remain in the weak strain zone.

In short, the ductile shear foliation and gneiss of this period may mainly occur in Archean, and they are the product of ductile deformation of the crustal rocks under a higher geothermal gradient. Due to the subsequent structural superimposition and deformation, the foliation of this period only remains in the weak strain domain.

2.5.2 Structure of Paleoproterozoic orogenic mélange belt

Paleoproterozoic orogenic belt structure is mainly recorded in the Huanglianghe Rock Formation and the Li'erping Rock Formation in the northern part of the Huangling Dome, including a series of tectonic deformations related to orogeny, i. e., ductile shear zone, NNE–NE-trending folds, NE-trending gneiss, and subsequent post orogenic extensional detachment structures.

In recent years, field geological surveys and studies on the pre-Nanhuaian basement in the northern part of the Huangling anticline in the Yangtze Craton have shown that it can be divided into three different geological tectonic units: the Shuiyuesi microcontinent in the west, the Bashansi microcontinent in the east, and the Kongling metamorphic complexes in between (which is the Palaeoproterozoic Kongling mélange belt). The Shuiyuesi microcontinent in the west is mainly composed of the Dongchonghe TTG gneiss of 2.95–2.90 Ga in Meso archean (高山等, 1990; 2001; Qiu et al., 2000; Zhang et al., 2006; 魏君奇等, 2009) and various-sized plagioclase amphibolite inclusions of 3.05–3.00 Ga in Meso archean(富公勤等, 1993; 魏君奇等, 2012, 2013). The Bashansi micro-continent in the east is mainly composed of Paleoproterozoic (2.33–2.17 Ga) granitic gneiss complex (biotite plagioclase granitic gneiss, biotite monzonitic granitic gneiss, etc.), as well as plagioclase amphibolite, biotite plagioclase gneiss, and other inclusions (姜继圣, 1986; 李福喜等, 1987). New research shows that the Bashansi microcontinent has the Paleoproterozoic granite gneiss of 2.33–2.17 Ga, the Neoarchean A-type granite gneiss of 2.7–2.6 Ga (Chen et al., 2013), as well as so far the oldest Paleoarchean TTG gneisses of 3.45–3.30 Ga in South China (Guo et al., 2014).

Therefore, whether formed in Paleoproterozoic, Neoarchean, or Paleoarchean, the main body of the Bashansi microcontinent in the northern part of the Huangling Dome is obviously different from the formation age and evolution history of the Mesoarchean Shuiyuesi microcontinent. This also indicates that there should be a collisional orogenic belt between the Shuiyuesi microcontinent and the Bashansi microcontinent, which is suggested as the Kongling mélange belt formed by the Paleoproterozoic subduction-collision orogen. However, the composition of blocks (slices) and matrix structure, temporal and spatial distribution, and genetic evolution characteristics in the orogenic mélange belt remains to be further studied.

2.5.2.1 *Ductile shear deformation structure*

Several important lithological interfaces in the study area are the basis for the development of ductile shear zones in this period, such as the interface between the Huanglianghe Rock Formation and the Dongchonghe gneiss complex, and the interface between the Huanglianghe Rock Formation and the Li'erping Rock Formation, which are well developed in Huanglianghe Rock Formation and Li'erping Rock Formation. The ductile shear deformation structure takes the lithological interface with large differences in primary bedding or competence as the deformation surface, which is mainly manifested as ductile shear deformation zone, as well as a number of rootless shear folds, viscous boudinage, and structural lenses.

Research shows that in the Huanglianghe Forest Farm, the contact surface between the Huanglianghe Rock Formation and the Dongchonghe gneiss complex is semi-annular, and the gneissic foliation is inclined to the southeast (outside). Along the contact surface, a nearly east–west ductile shear zone with a width of about 7 m is also developed, consisting of a 3 m-wide primary mylonite zone and a 4 m-wide shear fold zone, with granitic and biotite plagioclase mylonites developed. Stretch lineation and rotational porphyroclast system show early dextral bedding-parallel thrusting and late dextral bedding slip (Figure 2-16). The bedding-parallel ductile shear zone developed at the interface between the Huanglianghe Rock Formation and the Dongchonghe gneiss complex was formed at the end of Paleoproterozoic, and may be of a near-horizontal detachment type. Later, it was transformed by the diapir uplift of the Quanyitang granite, and the occurrence was steepened (熊成云等, 2004).

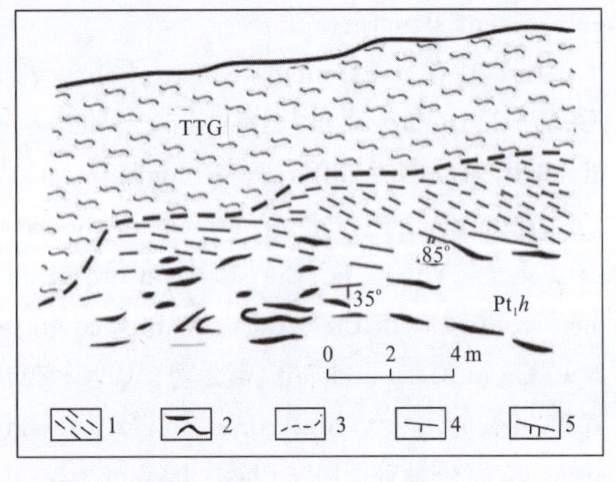

1. mylonite zone; 2. felsic zone; 3. lithological boundary; 4. gneissic foliation; 5. mylonite foliation.

Figure 2-16 Contact relationship between the Huanglianghe Rock Formation and the Dongchonghe gneiss complex (熊成云等, 2004)

In the Huanglianghe Rock Formation and the Li'erping Rock Formation, boudinage structures, tectonic lenses, plagioclase rotational porphyroclast system, and S-C fabrics are

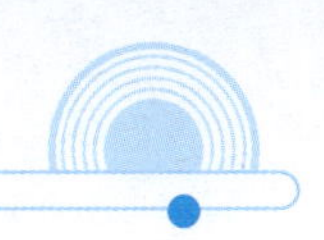

widely developed, which occur in bedding ductile shear zone, indicating that the rocks are subjected to strong vertical flattening (Figures 2-17 and 2-18). Macroscopically, a large number of marble structural lenses are seen in Erlangmiao, Qinjiahe, and other places in the Huangling basement. The long axis of the lenses is parallel to the gneissic foliation, and they are distributed intermittently in the region. They have experienced strong structural replacement, and generally strike NNE–NE.

Figure 2-17 The tectonic rotation porphyroclast developed in the TTG gneiss in the Wuduhe–Yinjiaping Highway section
(felsic and dark minerals distributed in strips, and the felsic rotation porphyroclast showing sinistral characteristics; photo by Han Qingsen)

Figure 2-18 TTG gneiss on the Wuduhe–Yinjiaping Highway section
(formed by ductile shear, with sinistral characteristics; photo by Han Qingsen)

2.5.2.2 *Penetrating gneissic structure*

The metamorphic rock series in the northern part of the Huangling Dome has extensively developed a group of NE-trending penetrating new-born gneiss (schistosity) (130°–180°∠40°), which is formed synchronously with the ductile shear zone. The structural deformation foliation in the Huanglianghe Rock Formation and the Li'erping Rock Formation are mainly developed along the primary bedding (S_0) plane, which is represented by the obvious lithologic interfaces among gneiss, marble, and plagioclase amphibolite. While inside a single lithological layer, it has been replaced by strong ductile gneissic (or schistosity) foliations (Figure 2-19). Gneissic foliations are widely developed in rocks such as biotite plagioclase gneiss, plagioblend amphibole gneiss, and biotite plagioclase gneiss. They are composed of plagioclase, hornblende, biotite, quartz, and other minerals in parallel orientation. The appearance of gneisses is banded, while the schistosity are mainly developed in marble and plagioclase amphibolite, which are composed of calcite or amphibole and other minerals in parallel orientation, forming stripes or striated structure.

Figure 2-19 Striped pomegranate biotite plagioclase gneiss, trending NE with a width of about several millimeters to several centimeters (the section of the Wuduhe–Yinjiaping Highway outcrops; photo by Han Qingsen)

The regional gneisses are affected by the later ductile shear zone. To the north of the Wuduhe fault zone, the foliation striking NE transitions southwardly (Yinjiaping–Erlangmiao–Maliangping) to NEE direction, approximately from east to west, and even northwest direction, forming an arc-shaped structure convex to the southeast uplift, with dip angle generally less than 50°. To the south of the Wuduhe fault zone, the foliation strike is nearly EW–NWW, and the foliation dip angle varies greatly, generally greater than 50°. This may be due to a joint effect of the basement emplacement and structural deformation of the Huangling granite in Neoproterozoic.

2.5.2.3 *NNE–NE fold structure*

NNE–NE folds on outcrop scale are common in the Wuduhe–Yinjiaping section of the high-grade metamorphic rock area in the northern part of the Huangling Dome. On the regional scale, the NE-trending folds include the Quanyitang dome-shaped composite antiform, the Bashansi composite synform, and the Baizhuping antiform. Xiong Chengyun et al. (2004) made a geological structure profile along SE120° (Figure 2-20). The axial trace of the Quanyitang dome-shaped complex antiform trends NE30°, which is the product of the transformation of the dome. The Quanyitang anitform is inclined to the southeast and secondary folds are well developed in the southeast limb. The latter two are inverted to the northwest direction, and the Bashansi composite synform trends NE25° with S-shaped axes, consisting of Bashansi synform, Hengdeng synform, and Guanmiao antiform. Three phases of superimposed folds can be identified on the NW limb; the axis trace of Baizhuping antiform is NE25°–30° and the core is basically occupied by Bashansi gneiss complex.

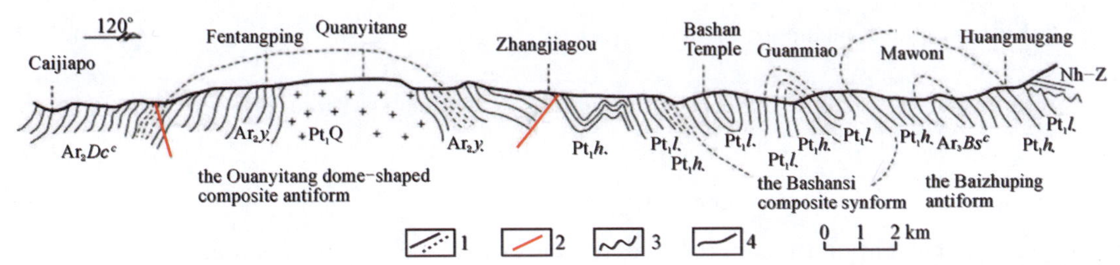

1. ductile shear zone; 2. fault; 3. unconformity; Ar_3Bs^c. the Neoarchean Dongchonghe gneiss complex; Ar_2Dc^c. the Mesoarchean Dongchonghe gneiss complex; Ar_2y. . the Mesoarchean Yemadong Rock Formation; Pt_1Q. Quanyitang unit of Paleoproterozoic; Pt_1h. . the Paleoproterozoic Huanglianghe Rock Formation; Pt_1l. . the Paleoproterozoic Li′erping Rock Formation; Nh–Z. Nanhuanian–Sinian.

Figure 2-20 Structural profile of the Huangling Dome core

(熊成云等，2004)

2.5.2.4 *Extensional deformation structure*

1. Basic diabase dykes (veins)

The common basic diabase veins in the metamorphic basement of the northern Huangling Dome occur as dykes. The strike of diabase veins is mainly NNW and NE, and the dip angle is close to 90°. The dykes have obvious intrusive relationship with the surrounding rocks, and occasionally have condensing edges. Some dykes contain gneiss xenolith (Figure 2-21). These basic rock veins are 0.4–3 m wide, and a few are above ten meters. The lithology is mainly diabase or gabbro, without obvious metamorphic deformation characteristics. Most diabase dykes developed two groups of nearly perpendicular joints due to the late tectonic force, resulting in the crushing of the rock (Figure 2-22). According to the statistical analysis of the occurrence of 40 basic dykes measured in the field, it is found that there are mainly two groups of preferred orientations: NW330°–340°, NE40°–50°. A large number of developed basic dykes show that they were formed in an extensional tectonic environment.

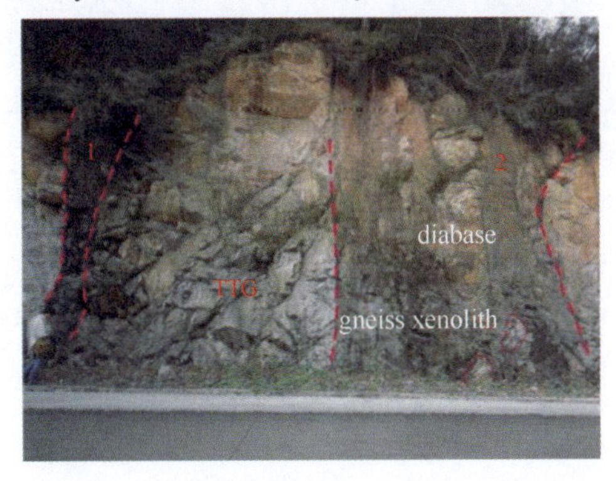

1. tectonic breccias of gneiss; 2. tectonic lens of gneiss.

Figure 2-21 Several diabase dykes intruded into gneiss near the Tandanghe (containing lenticular gneiss xenoliths; photo by Han Qingsen)

Figure 2-22 Diabase dykes intruded into gneiss near Gongjiahe (containing gneiss xenoliths; photo by Han Qingsen)

2. Extensional detachment structure

A large number of low angle bedding-parallel detachment structures are developed in the granite gneiss complex along the Wuduhe–Yinjiaping Highway section in the northern part of the Huangling Dome. The tectonic zone in the field show a Z-shaped feature, presenting an extension detachment on the outcrop scale (Figures 2-23 and 2-24). Combined with previous studies, it might be a product of the composite dome structure of the Quanyitang granite in the post-orogenic crustal uplift and extension stage. The Quanyitang dome may be a composite result of the Archaean Yemadong–Dongchonghe gneissic doming, the Quanyitang superimposed fold uplift, and the Quanyitang K-feldspar granite doming.

Figure 2-23 A series of nearly EW detachment structures developed in TTG gneiss near the Tandanghe of Wuduhe–Yinjiaping Highway, with steep dip and normal faulting (photo by Han Qingsen)

Figure 2-24 Z-shaped detachment structure developed in biotite plagioclase gneiss near the Tandanghe of Wuduhe–Yinjiaping Highway, with nearly vertical diabase dyke intrusion on the left side (photo by Han Qingsen)

2.5.3 Structure of Neoproterozoic ophiolitic mélange belt

The Neoproterozoic ophiolitic mélange zone in the Huangling Dome area is represented by the Miaowan ophiolitic mélange distributed between Taipingxi and Dengcun. This set of ophiolitic mélange has undergone strong ductile and brittle deformation and metamorphism, and superimposed by folds. The overall trend of the ophiolitic mélange is NWW, the foliation is nearly vertical dipping generally northward, and it occurs as parallel belts (Figure 2-25).

2.5.3.1 *Ductile shear deformation structure*

The ductile shear deformation structure is mainly exposed in the Meizhichang and Maoya areas. Together with the regional foliation (gneisses) structure, it constitutes the early

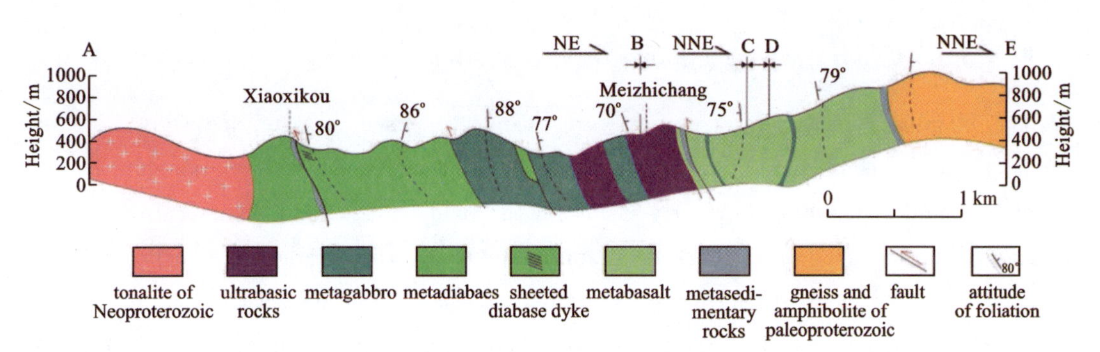

Figure 2-25 Geological profile of the Miaowan ophiolite in the southern part of the Huangling Dome (彭松柏等, 2010; Peng et al., 2012)

deformation characteristics of the Miaowan ophiolitic mélange, causing the rock units in the mélange zone to suffer at least up to regional deformation and metamorphism of high amphibolite facies. The serpentinite, serpentinized peridotite, and harzburgite that experienced ductile shear deformation in the early stage were transformed by extensional deformation to fracture zones, as well as serpentinized peridotite and pyroxene peridotite lenses (Figure 2-26). The attitude statistics of the structural fracture zone show a preferred orientation of 65°∠75°. The early ductile shear caused the formation of a large number of structural differentiation veins and new penetrating foliations in the layered basalt (Figure 2-27). These penetrating ductile deformation foliations are mainly manifested by the oriented arrangement of hornblende (formed by pyroxene degeneration) and plagioclase. The representative orientation of ductile deformation foliation is 47°∠79°.

Figure 2-26 Ultrabasic rock fracture zone in Meizhichang (photo by Jiang Xingfu)

Figure 2-27 Early penetrating foliation in metamorphosed basalt from North Meizhichang (photo by Jiang Xingfu)

2.5.3.2 *Meta diabase-gabbro intrusive structure*

The magmatic intrusive structures are mainly exposed in the areas of Xiaoxikou (Manshuiqiao), Yuanzifen, and Gucunping. The intrusive contact relationship between metadiabase and metagabbro is clearly visible in the field. Subsequent structural deformation and transformation are weak, but the amphibolite facies retrograde metamorphism occurs locally. Figure 2-28 shows the intrusive contact relationship between metadiabase and metagabbro. Figure 2-29 shows metadiabase intrusions into metagabbros, and the edge of the metadiabase has a quick crystallization due to the sudden temperature drop. Minerals such as hornblende and plagioclase are small in sizes, and the metadiabase crystal particles far away from the contact zone of the two are relatively coarse.

Figure 2-28 Metadiabase intruding into metagabbro at Xiaoxikou, Manshuiqiao (photo by Jiang Xingfu)

Figure 2-29 Condensation edge structure developed in metadiabase at Xiaoxikou, Manshuiqiao (photo by Han Qingsen)

2.5.3.3 *Metadiabase sheeted dyke*

The metadiabase sheeted dyke is only found in the Xiaoxikou area, with a length of about 450 m. The width of the sheeted dyke ranges from several centimeters to several meters, but most of them are 30–50 cm wide. The lithology is mainly metadiabase, followed by metagabbro and metaplagiogranite. The strike of the sheeted dyke is NWW with dip angles of 70°–80°. The sheeted dyke experienced strong deformation and metamorphism, with metamorphic grade of amphibolite facies. Most of the diabase veins in the sheeted dyke have bidirectional chilled margins, and a few have a single chilled margin (Figures 2-30 and 2-31). This is also an important evidence for the formation in an expansion of ocean floor (Deng et al., 2012).

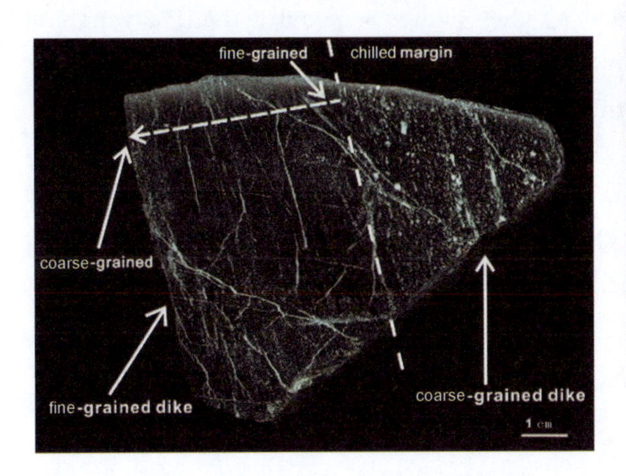

Figure 2-30 Unidirectional chilled margin of Xiaoxikou metadiabase (Deng et al., 2012)

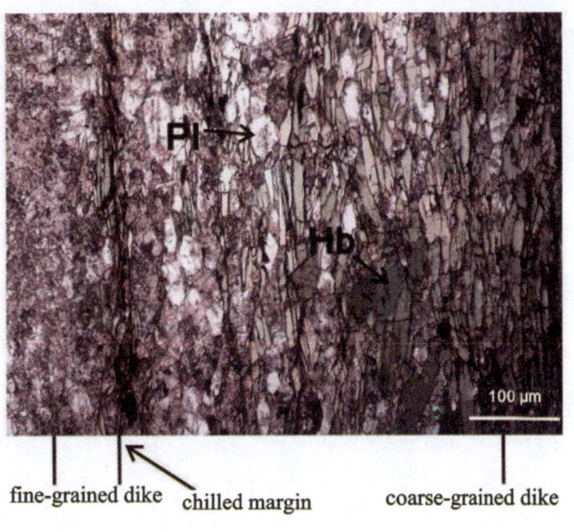

Figure 2-31 Unidirectional chilled margin of Xiaoxikou metadiabase (single polarized light) (Deng et al., 2012)

2.5.3.4 *Thrust fault structure*

This structure is mainly exposed in Xiaoxikou, located on the south side of the Miaowan ophiolitic mélange, and mainly developed in mica schist and metamorphosed sandstone. The overall strike of the fault is northwest, and the dip angle is generally greater than 60° (Figure 2-32). The "step" structure developed on the fault contact surface and drag structure show a thrust sense. The rock alteration in and near the fault zone is intense, mainly manifested as epidote and mica flake. Structural deformation phenomena such as secondary fold deformation, joints, and bedding schistosity can be seen in the rock layers on both sides of the fault.

Figure 2-32 High-angle thrust shear fault at Xiaoxikou

2.5.3.5 *Mesozoic–Cenozoic extensional metamorphic core complex structure*

The axis of the Huangling Dome is NNE, and the ratio of the long and short axis is about two to one. The periphery is bounded by the Xiannvshan fault, the Tianyangping fault, the Tongchenghe fault, and the Xinhua fault. From a regional perspective, the east and west sides of the Huangling Dome are the Jingmen-Dangyang basin and the Zigui basin, which form an obvious uplift-depression structure with the surrounding basins. Jiang Linsheng et al. (2002) believed that the deformation of the basement and sedimentary cover of the Huangling Dome has the characteristics of metamorphic core complex. Field observation and structural geometry analysis show that the two limbs of the Huangling Dome are steep in the west and gentle in the east, forming an asymmetric short-axis anticline dome structure, as shown in Figure 2-33 (王军等, 2010; Ji et al., 2013).

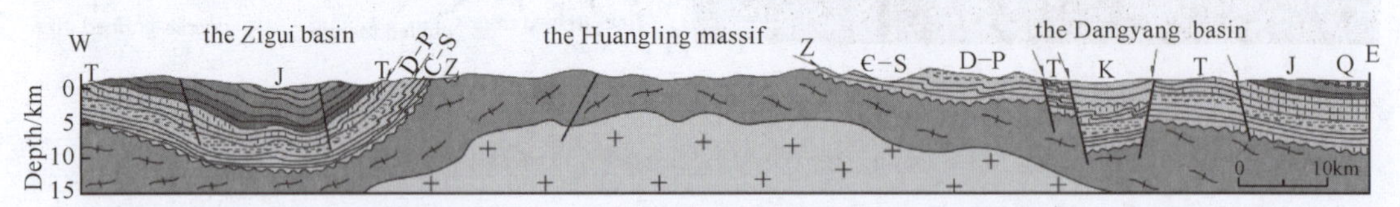

Figure 2-33 Structural profile of the Huangling Dome

(Ji et al., 2013)

There has always been controversy about the formation time of the Huangling Dome structure. Some researchers believe that the dome structure was formed in Neoproterozoic, or in Early Paleozoic, Early Mesozoic, and Late Mesozoic (江麟生等, 2002; 李益龙, 2007). However, field observations and studies in recent years have shown that the upper Jurassic in the western part of the Huangling Dome are obviously involved in deformation. Similarly, the Early Cretaceous sedimentary basins developed on the southwest and southeast limbs of the dome are obviously unconformable with the Huangling Dome observed today. Moreover, a large number of thermochronological studies have also shown that the uplift of the Huangling Dome mainly occurred between 110 Ma and 160 Ma (刘海军等, 2009; 沈传波等, 2009; Ji et al., 2013), and the Huangling Dome is still in the stage of uplift in Cenozoic (郑月蓉等, 2010; 肖虹等, 2010). Therefore, the Huangling Dome structure was mainly formed between Late Jurassic and Cretaceous, which is consistent with the tectonic dynamic background of Mesozoic–Cenozoic lithosphere extension and thinning in eastern China.

In addition, the Mesozoic–Cenozoic extensional structure of the Huangling Dome is mainly characterized by the deformation of cover rocks, the development of bedding detachment folds, extensional cataclastic lenses, and high angle normal faults, with local small-scale sliding thrusts. It is widely developed in the Early Triassic thin limestone, the Silurian shale of the Longmaxi Formation, the Ordovician limestone, the Cambrian carbonaceous limestone, especially the Sinian thin-bedded limestone of the Doushantuo Formation. The lithotectonic lens formed by stretching and breaking has obvious characteristics of vertical thinning and gravitational stretching (Figures 2-34 and 2-35).

Therefore, the Huangling Dome is mainly an extensional structure developed in Mesozoic–Cenozoic, which can also be called an extensional metamorphic core complex structure (Davis et al., 2002; 江麟生等, 2002; 肖虹等, 2010).

Figure 2-34 The lenticular siliceous rock in the black siliceous shale of the Doushantuo Formation in the northeast limb of the Huangling Dome (photo by Peng Songbai)

Figure 2-35 Fractured lens of argillaceous dolomite formed by bedding extension and detachment in the Doushantuo Formation (Jiuqunao) in the southwest limb of the Huangling Dome (photo by Peng Songbai)

2.5.3.6 *Major large-scale ductile-brittle fault structures*

Fractue structures are widely developed in the core and peripheral areas of the Huangling Dome, including ductile shear zones and brittle faults. There are mainly three groups of faults in nearly EW, NW, and NE directions, and the NW-trending ductile-brittle fault zone is the most prominent. The near EW-trending ductile shear zone is mainly composed of Shuiyuesi–Baizhuping fault zone in the northern part of the core, characterized by thrusts and slips, with slipping generally following thrusting. The NE-trending ductile shear zone is small in scale and is characterized by strike-slip and thrust. The NW-trending ductile-brittle shear zone is the most well developed, represented by Wuduhe, Bancanghe, and Dengcun–Xiaoxikou ductile-brittle fault zones. The late brittle fault activity is superimposed on the early ductile shear zone. Generally, left-lateral thrusting was followed by right-lateral slipping. It has a long activity history and is superimposed by multi-phase structural deformation. It is also the controlling structure of gold ore in the area (熊成云等, 1998). The basic characteristics are briefly described as follows.

1. The NNW-trending Xiannvshan fault zone

Located in the southwest of the Huangling Dome, it almost obliquely cuts all major east–

west folds in the survey area. It is a fault zone composed of a series of pinnately arranged faults. It is 80 km long, with an overall strike of NW20°, dipping to southwest with a dip angle of 40°–60°, cutting through the Paleozoic–Cretaceous strata (Figure 2-36). In some areas, the Paleozoic strata are thrusted over the Cretaceous strata. The compression of the fault zone is obvious. The fault breccia is developed. The breccia is angular and mixed in size, generally 1–5 cm. It has the nature of tensile breccia, and there are also mylonite belts and tectonic lenses with the nature of compression. Calcite veins are well developed in the zone, and the slicken sides are developed, most of which are horizontal, and a few dip about 10°. The strata on both sides of the fault are dislocated, and a large number of drag structures can be seen.

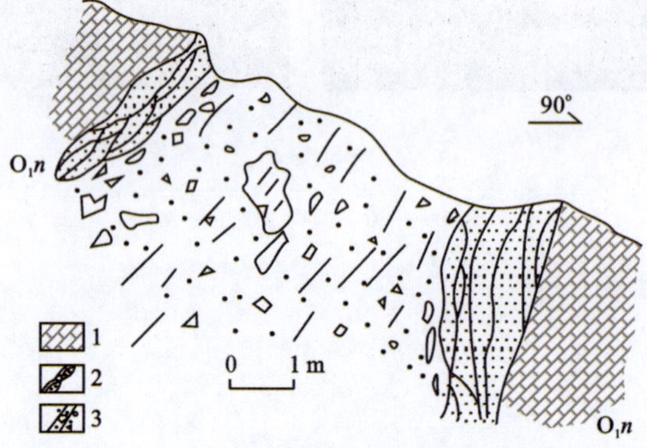

1. limestone of the Nanjinguan Formation; 2. compressive tectonic lens; 3. fracture zone.

Figure 2-36 Section of the Xiannvshan fault structure

(王辉等, 2010)

The fault activity is obvious. Along the fault, the negative topography is arranged in a linear pattern. Unconsolidated fault gouge and fault breccia can be seen in the fault zone. The valleys on both sides have obvious changes. The modern collapse and landslide develop. A famous one is the Xintan Giant Compound Landslide in 1985. There are not many large earthquakes along the fault, while microseisms are continuous. However, due to the complex geological conditions, various geological hazards along the fault are well developed.

2. The Near NS-trending Zhouping fault zone

It is located on the west side of the Huangling Dome, with a general trend in the NNE direction, and the southwest end merges with the Xiannvshan Fault. It may be a branch structure of the Xiannvshan Fault, with an exposed length of about 15 km. The fault trends NE20°, dips mainly ot the west with steep dip angles, most of which are above 70°. In some areas, the fault surface is almost vertical, and it can be seen that it cuts through the Holocene strata in the middle and late Quaternary. The fault has a certain degree of activity. Along the fault zone, the negative topography develops, the stream system on both sides changes

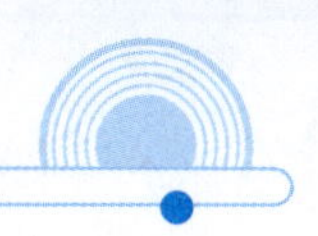

obviously, and there is a certain degree of microseismic activity. The surface deformation monitoring shows that the fault zone has differential activities.

3. The NW-trending Bancanghe fault zone

The Bancanghe fault zone is a regional large fault with an overall strike of about NW310°. It extends from the upper Qianyanghe, through the Bancanghe to Hongjiaping, extending northwest and southeast beyond the survey area, with a length of 16.95 km. The main fault surface dips to northeast–north-northeast with a dip angle of 60°–78°. In the Bancanghe and Sunjiahe areas, the fault dips to the southwest with a dip angle of 50°–70°. The width of the fault fracture zone is generally less than 30 m and locally up to 55 m.

The metamorphic basement area is mainly composed of cataclastic, granulite, porphyry, fault (breccia), cataclastic granite, and cataclase diorite of different stages, and most of them occur as structural lens. The fault (angular) breccia is composed of granitic mylonite and mylonitized granite. Cleavage zones are well developed, parallel to the fault. In the fault fracture zone of the cap area, a large number of ductile shear structural lenses (composed of carbonite fault breccia) were developed in the early stage and structural breccia ware developed in the late stage, which also show the characteristics of ductile shear and compression in the early stage and the subsequent brittle extension in the late stage.

4. The NW-trending Wuduhe fault zone

The Wuduhe fault zone is a regional large fault with a strike of NW320°–330°, passing through the northern part of the field practice area and spreading along Guanyintang–Wuduhe River–Huamiao. The outcrop is 37 km long, extending into the sedimentary cap along the northwest and southeast directions respectively. The faults mainly cut through the metamorphic rocks, and some sections cut through the Chalukou superunit and the Sinian strata. The overall dipping of the faults in the region is mainly northeast, and in the basement area is mainly southwest, with a dip angle of 62°–87°. The fault zone in Precambrian basement has a width of more than 50 m, which is mainly composed of fault rocks, granulite, porphyry, mylonitic fault conglomerate, and fault breccia of different stages. The fracture zone of the cap area is about 10–20 m in width, which is mainly composed of fault breccia, cataclasite, etc. It is a large ductile brittle fault zone with long-term activity.

A series of roughly parallel secondary fault surfaces and cleavage are common in this fault zone, either straight or in a gentle wave shape, with multi-phase activities. The early stage belongs to ductile shear deformation fault zone, and the late stage is characterized by brittle deformation activity. The early and middle stages of brittle deformation are dominated by thrusting with strike slipping, and the late stage is a strike-slip fault with

normal faulting. The fault fracture zone in the basement area is characterized by silicification, epidotization, limonitization, pyritization, Pb and Zn mineralization, etc. The late diabase (porphyritc) vein, diorite (porphyrite) vein, granite vein, and biotite monzogranite vein are commonly distributed along the fault zone. The regional foliation strike on both sides of the fault is similar, and the common red granitic veins are distributed along the wall rock foliation, which indicates that the fault was formed at least in the early and middle Neoproterozoic.

This fault zone is also an important ore-conducting and ore-controlling structure for gold, molybdenite, magnetite, and rare radioactive minerals in this area. The pyrite mineralization is closely related to the late tensional shear deformation. Gold deposits mainly occur in the NNW–NW secondary fault zone adjacent the fault zone. During Yanshanian, there was inherited brittle fault activity, which cut into the cap rock. The fault occurs as a linear feature on aerial photos, and the topography is mostly negative (passes, straight stream ditches, etc.), and the fault triangle facets are well developed in the area of Guanyintang–Maopinghe–Chalukou. According to the 1/200,000 regional survey, the fault is brittle in the late stage, and cuts through the Cretaceous, which indicates that its activity lasts at least until after Cretaceous.

2.5.3.7 *Regional geological structure evolution*

The Huangling Dome is located in the core area of the Yangtze Craton, where the oldest Archean gneiss complex (TTG) in South China and the Paleoproterozoic hornblende-granulite facies high-grade metamorphic complex are exposed. It is an important window for studying the early geological tectonic evolution of South China, and the assemblages and breakups of Precambrian Columbia and Rodinia supercontinents, recording multi-phase important subduction/accretion-collision orogeny-collage events. In particular, the pre-Nanhuaian metamorphic magmatic complex in the crystalline basement in the core of the Huangling Dome is a relatively complete record, which preserved the growth evidences of the Archean paleo-continental crust, the Paleoproterozoic subduction-collision orogeny (high pressure granulite, etc.), the Neoproterozoic subduction-collision orogeny and rifting (ophiolite, granite complex, etc.), the Mesozoic–Cenozoic Huangling Dome uplift, extension, and thinning (metamorphic core complex structure), and other important geological events. The regional geological tectonic evolution in Huangling area can be roughly divided into two important tectonic evolution stages, basement and cap stages, which are briefly described as follows.

1. Tectonic evolution stage of the basement

1) Formation of the Archean paleo-continent nucleus (microcontinent block)

The Archean continental crust in the core of the Huangling Dome is characterized by the

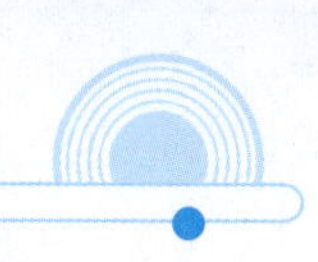

formation of the granite-greenstone terrane paleo-continent nucleus (microcontinent) of the Yangtze Craton. The Archean Dongchonghe granite gneiss system (TTG) and the amphibolite of the Yemadong Rock Formation are its main material records. The emplacement age of the Archean granite gneiss (TTG) is 3450–2900 Ma (高山等, 1990; Gao, et al., 2011), which is an important product of early continental crust formation, evolution, and growth.

2) Paleoproterozoic subduction/accretion-collision orogeng-collage

In the early Paleoproterozoic, the northern part of the Huangling Dome was characterized by sedimentary combination: Superior-type iron formations (banded iron formations, BIFs) composed of quartzite, ferrous rock, and schist, and formations of terrigenous clastic claystone with higher maturity, siltstone intercalated with carbonate, siliceous rock, and carbonaceous mudstone, and clastic carbonate (the khondalite formation). The volcanism was weak. In the late Paleoproterozoic, the northern part of the Huangling Dome entered the tectonic evolution stage of subduction and collision orogeny, characterized by the formation of NNE–NEE-trending amphibolite facies-granulite facies metamorphic belt (2.0–1.95 Ga), and A-type granite, subvolcanic volcanic rocks, and basic dykes (about 1.85 Ga) under the post orogenic extension system. It indicates that an important tectonic event took place in the northern area of the Huangling Dome at 2.0–1.85 Ga. The tectonic event from subduction collision orogeny to post orogenic extensional collapse may be related to the global assemblage and breakup of the Columbia supercontinent (凌文黎, 1998; Zhang et al., 2006b; 熊庆等, 2008; Wu et al., 2009; Cen et al., 2012; Yin et al., 2013; Peng et al., 2012).

3) Meso–Neoproterozoic subduction-collision orogeny-collage

The discovery of the Neoproterozoic Miaowan ophiolite in the south of the Huangling Dome indicates that the Yangtze Craton basement was consolidated and formed by several distinctive blocks and terrains through Neoproterozoic subduction collision orogeny (i. e., the Greenwell Movement) to form the basic outline of the Yangtze Craton basement (彭松柏等, 2010; Peng, et al., 2012). In the early Neoproterozoic (960–870 Ma), the Shennongjia Island Arc and the Yangtze Block had a subduction-collision orogeny-collage, which led to the tectonic emplacement of the Miaowan ophiolitic mélange. The late Neoproterozoic (860–790 Ma) subduction-collision orogeny extension collapsed tectonic environment formed the adakitic/island arc volcanic granite—namely the Huangling granite complex (Zhang et al., 2008; Wei et al., 2012; Zhao et al., 2013). About 790 Ma, the Yangtze Craton and the basement tectonic evolution stage of this area ended, the tectonic movement was dominated by differential uplift and subsidence, and it entered the stable cover sedimentary evolution stage.

2. Tectonic evolution stage of the cap

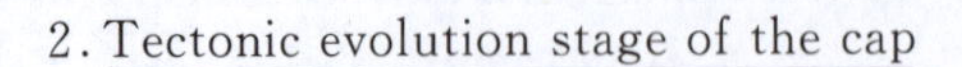

1) Nanhuaian–Early Mesozoic marine stable sediments

During this period, the Yangtze Craton deposited a set of terrigenous clastic sediments of meandering river-estuarine delta branch channels on the basis of uplift and denudation, and subsequently deposited the continental glacial sediments of the Nantuo Formation, which is also the result of the global "snowball Earth" event. After the Doushantuoian, a set of stable cratonic marine sedimentary rocks mainly composed of black shale and carbonate rock from basin margin facies to restricted platform facies was continuously deposited. The tectonic uplift began to occur in Early Mesozoic and Late Triassic under the influence of the Indosinian movements (沈传波等, 2009).

2) Late Mesozoic–Cenozoic intracontinental compression-extension deformation

Since Late Mesozoic, the Huangling Dome and its peripheral areas have entered the period of intracontinental compression-extension tectonic evolution. The compression and deformation of the Indosinian–Early Yanshanian movements formed the development of Jura-typed folds in the cover. Since Late Yanshanian, the lithosphere has undergone strong tectonic uplifting under the influence of strong extensional thinning, forming the basic prototype of the Huangling Dome. The horizontal detachment fold and high-angle extension brittle normal faults have been formed in the cover sedimentary strata. The contact zone between the sedimentary strata and the basement developed low-angle bedding cleavage and ductile shear faults (沈传波等, 2009; 刘海军等, 2009; Ji et al., 2013), accompanied by magmatic hydrothermal ore-forming activities, laying the basic outline of the metamorphic core complex structure of the Huangling Dome.

The Cenozoic in eastern China and the Huangling Dome area is mainly controlled and influenced by the Himalayan movements, the uplift of the Tibetan Plateau, and the subduction of the Pacific Plate. It mainly manifested as intermittent tectonic uplift under the combined action of the compression-extension tectonic system (陈文等, 2006; 李海兵等, 2008; 郑月蓉等, 2010; 肖虹等, 2010). The Three Gorges area of the Yangtze River has intense incision, forming multi-level structural terraces, high mountains and deep valleys, slopes and cliffs, and karst development, as well as frequent geological hazards like landslides and rock falls (谢明, 1990; 李长安等, 1999).

II

FIELDTRIP ROUTES

3 ROUTE ONE: OBSERVING THE NEOPROTEROZOIC NANHUAIAN PERIOD STRATIGRAPHIC SEQUENCE

3.1 Teaching Route

Zigui Base–Gaojiaxi, Sandouping Village–Huajipo Village–Jiulongwan–Huangniuya–Zigui Base

3.2 Teaching Tasks and Requirements

(1) Observing and describing the lithology of the Liantuo Formation, the Nantuo Formation, the Doushantuo Formation, and the bottom of the Dengying Formation, as well as those corresponding sedimentary structures, depositional environment, and stratigraphic contact relationships.

(2) Understanding the characteristics of paleogeography, paleoclimate, and paleoecology during Neoproterozoic Nanhuaian (Cryogenian) and Sinian (Ediacaran).

(3) Sketching the crude stratigraphic column (1/5,000).

(4) Collecting some typical rock and mineral samples.

(5) Probing into some scientific questions and focused studies.

3.3 Route Information and Observing Points

The well-exposed Neoproterozoic strata represents typical stratigraphic units in the Three Gorges region. Owing to its good continuity, it is commonly applied as the standard stratigraphic sequence for the Neoproterozoic strata in South China (Table 3-1).

3.3.1 The Liantuo Formation (Nh_1l)

The Liantuo Formation is featured by fuchsia thick-bedded conglomerate, fuchsia medium thick-bedded gravel-bearing quartz sandstone, fuchsia feldspathic quartz sandstone, fuchsia medium-bedded medium-grained sandstone, fuchsia thin-bedded siltstone, fuchsia silty shale, grayish-green shale, mudstone, etc.

The Huangling massif is unconformably overlain by the Liantuo Formation (Figure 3-1, left). The observation site is located behind a civilian house near the bridge in Gaojiaxi (GPS: N30°46′19.9″, E111°01′9.6″). Evidence includes:

(1) The top surface of Huangling massif is eroded, which is also indicated by the paleosoil.

(2) The bottom of the overlying Liantuo Formation consists of several layers of conglomerate which contains some gravel initially belonging to the Huangling pluton.

(3) There is a gap between those two units. The underlying Huangling massif is dated 800 Ma or so while the overlying the Liantuo Formation is confined within 750 Ma. The sedimentary environment of the Liantuo Formation can be interpreted as diluvial fan, alluvial fan, delta, and littoral facies in an ascending order. Beyond that, large cross-bedding (Figure 3-1, right) and some authigenic minerals (e.g., glauconite) can be found in the sandstone. Regionally, the thickness of the Liantuo Formation is about 190 m.

3.3.2 The Nantuo Formation (Nh_1n)

The Nantuo Formation is dominated by grayish-green or grayish-purple massive anagenite, gravel-bearing sandy mudstone and siltstone intercalated with some thick–medium-bedded gravel-bearing muddy tillites. The tillite is generally grayish-green and non-bedded, with varied-sized till cobble on which some glacial T-shaped striae were discovered. All the above were considered as material records during the glacial period.

Table 3-1 Neoproterozoic Nanhuaian and Sinian stratigraphic sequence in Zigui, Hubei

Time				Formation	Member	Code	Thickness/m	Lithological Description
Neoproterozoic	Sinian	Upper	Longdengxi	the Dengying Formation	Baimatuo	Z_2dy^b	17.50	grey thick–medium-bedded dolomite, intercalated with medium–thin-bedded fine crystalline dolomite, with partial siliceous banding and nodules
			Shibantan		Shibantan	Z_2dy^s	36.0	dark gray and greyish-black thin-bedded micritic limestone, with some chert banding and very thin-bedded banding of micritic dolomite
			Hamajing		Hamajing	Z_2dy^h	134.4	grey–light grey medium–thick-bedded intraclast dolomite, fine crystalline dolomite, fine crystalline siliceous dolomite
		Lower	Miaohejie	the Doushantuo Formation	4th	Z_1d^4	44.1	black thin-bedded siliceous mudstone and carbonaceous mudstone intercalated with dolomitic limestone
					3rd	Z_1d^3	60.9	The top is grey thick–medium-bedded dolomite, powder–fine crystalline dolomite, with chert nodules and banding. The bottom is thin-bedded crystalline powder dolomite
	Nanhuaian		Weng'anjie		2nd	Z_1d^2	89.2	dark grey–black thin-bedded argillaceous limestone and dolomite intercalated with thin-bedded carbonaceous mudstone, unevenly layered
					1st	Z_1d^1	5.5	grey and dark greyish-black thick-bedded siliceous dolomite, with chert nodules; thin–medium-bedded dolomite, lime dolomite
		Upper		the Nantuo Formation		Nh_2n	103.4	greyish-green with fuchsia massive glacial conglomerate and tillite, with partial thin-bedded silty mudstone occassionally seen
		Lower		the Liantuo Formation	2nd	Nh_1l^2	30	fuchsia thin–medium-bedded sandstone, interbedded with silty mudstone and siltstone, with some medium–thick-bedded gravel sandstone occassionally seen
					1st	Nh_1l^1	63	fuchsia and greyish-green thick-bedded feldspar quartz sandstone, gravel-bearing sandstone, and sandstone, intercalated with thin-bedded tuffaceous quartz sandstone and some thin-bedded silty mudstone
Mesoproterozoic				the Miaowan Formation		Pt_2m	864	amphibolite
Palaeoproterozoic				the Xiaoyucun Formation		Pt_1x	645	biotite plagioclase K-feldspar gneiss, plagioclase gneiss, quartz biotite schist, or amphibolite

Figure 3-1 The stratigraphic unconformity contact (left) and cross-bedding (right) of the Nanhuaian Liantuo Formation in Gaojiaxi

Unconformity between the Nantuo Formation and the underlying Liantuo Formation is obvious, while the Gucheng Formation and the Datangpo Formation are missing (Figure 3-2). It's worth noting that the contact relationship on the Jiulongwan section is likely to be deformed by faults.

Figure 3-2 Unconformity between the Nantuo Formation and the underlying Liantuo Formation (Qinglinkou)

The genesis of the tillite of the Nantuo Formation relates to the Neoproterozoic glaciation. It is concerned as the product of rapid accumulation when Marinoan was over. Having studied the relating stratigraphic sequences and conglomcratc components, some researchers propose that the tillite in the Nantuo Formation is not accumulated in a single phase. On the contrary, it may have experienced multi-cycles of ice transgression and regression (Hu et al., 2012; Figure 3-3) and the provenance of gravels may scatter on different ancient continents. The thickness of the Nantuo Formation varies from tens to hundreds of meters in different areas. The exposed tillite of the Nantuo Formation in the Three Gorges area has a thickness of about 100 m.

3.3.3 The Doushantuo Formation (Z_1d)

The Doushantuo Formation is segmented into four members. The first member is mainly gray thick-bedded dolomitic limestones, also known as "cap carbonates". The silicalite developed along the bedding, which is in parallel unconformity with the underlying tillite of the Nantuo Formation. The second member is composed of the dark gray thin-bedded argillaceous limestone and dolomite intercalated with black thin-bedded carbonaceous shale and mudstone. Besides, some centimeter-level phosphorite nodules which was reported to have contained embryo fossils of the early Earth are found within the strata. The third member is featured by off-white thick–thin-bedded dolomite intercalated with black stripped and crumby chert. The fourth member consists of black thin-bedded carbonaceous mudstone and siliceous mudstone with lenticular dolomitic limestone.

The Doushantuo Formation was the product of the snowball Earth event during sea-level rising periods. The exposed Doushantuo Formation in the Three Gorges area might indicate lagoon and marine shallow shelf. The bottom-up dramatic fluctuation of carbon and sulfur isotopes of the Doushantuo Formation is attributed to the frequent changes of its sedimentary environment (Figure 3-4). The thickness of this unit is about 150 m.

3.3.4 The Dengying Formation (Z_2dy)

The Dengying Formation is divided into three members from bottom to top: the Hamajing Member is characterized by light gray thick-bedded dolomite which is in conformable contact with the underlying Doushantuo Formation; the Shibantan Member consists of blackish-gray laminated micrite with chert bandings; the Baimatuo Member consists of grayish-white and light gray thick–thin-bedded dolomite with siliceous bandings.

The top part of the Dengying Formation is the Yanjiahe Member which is composed of grayish-yellow marlstone, carbonaceous limestone, and shale with siliceous bandings and nodules. This unit has been assigned to the bottom of Cambrian. Regionally, it is in conformable contact with black carbonaceous shale and limestone with carbonate concretions of the Cambrian Shuijingtuo Formation.

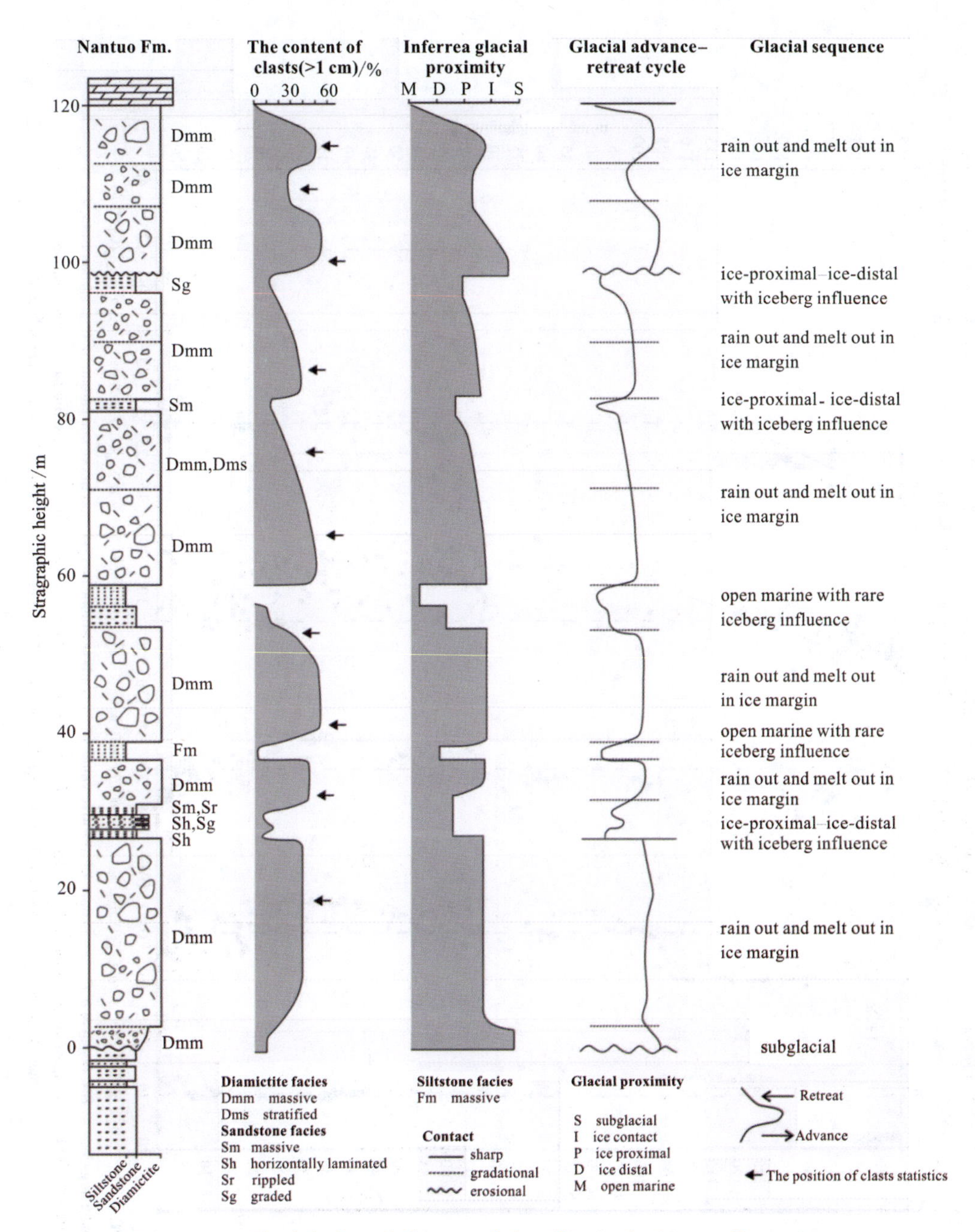

Figure 3-3 The lithological changes of the tillite in the Nantuo Formation at Jiulongwan and the corresponding indication of ice transgression and regression (Hu et al., 2012)

For the whole Dengying Formation, only the Hamajing and Shibantan Members are exposed on this teaching route. It can be discovered that the Dengying gray thick-bedded dolomite overlays the Doushantuo black carbonaceous shale and lenticular limestone (Figure 3-5). The thickness of the Dengying Formation tends to be larger, which is approximately 160 m.

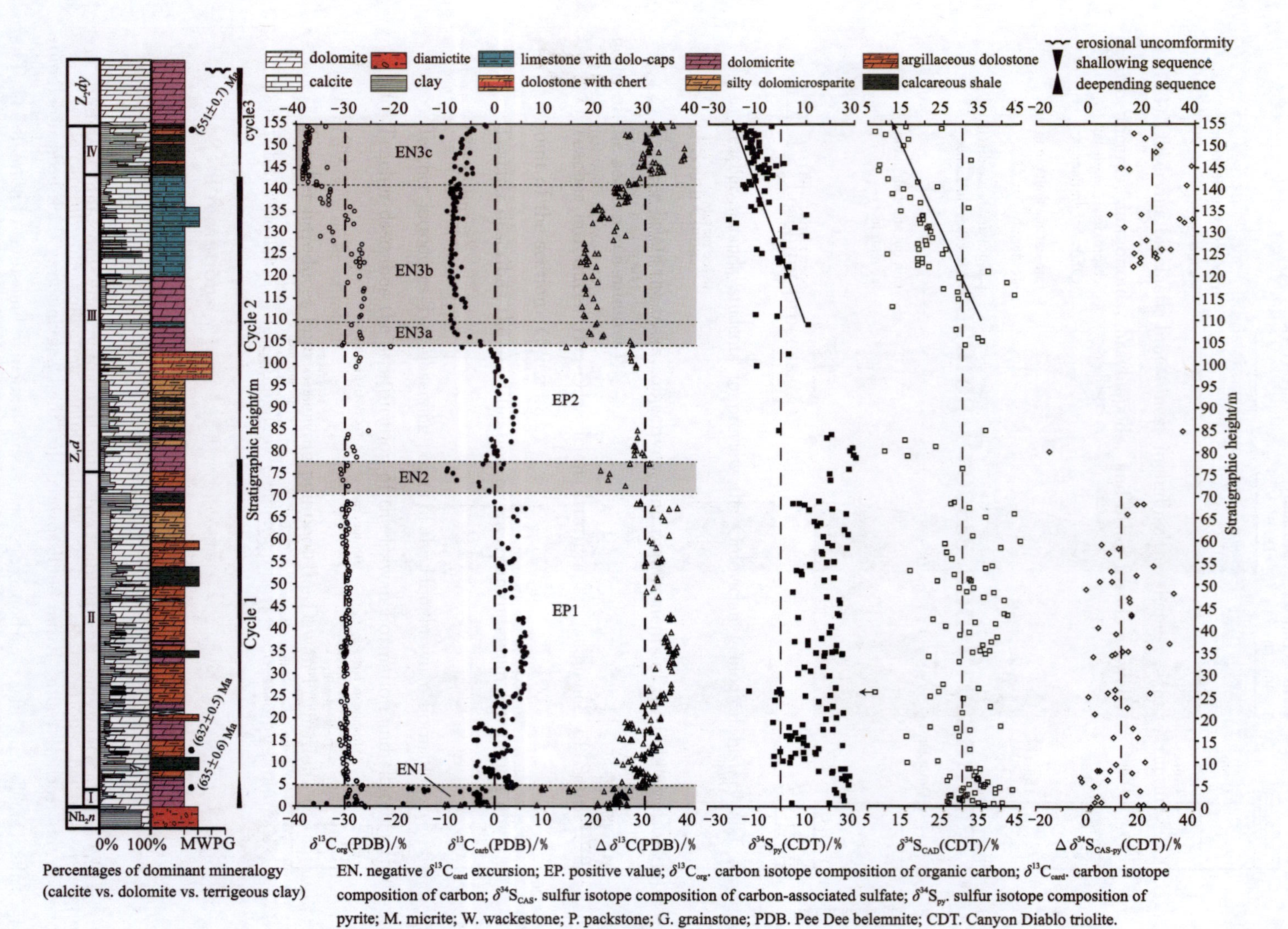

Percentages of dominant mineralogy (calcite vs. dolomite vs. terrigeous clay)

EN. negative $\delta^{13}C_{card}$ excursion; EP. positive value; $\delta^{13}C_{org}$. carbon isotope composition of organic carbon; $\delta^{13}C_{card}$. carbon isotope composition of carbon; $\delta^{34}S_{CAS}$. sulfur isotope composition of carbon-associated sulfate; $\delta^{34}S_{py}$. sulfur isotope composition of pyrite; M. micrite; W. wackestone; P. packstone; G. grainstone; PDB. Pee Dee belemnite; CDT. Canyon Diablo triolite.

Figure 3-4 Chemostratigrapnic profiles of the Doushantuo Formation at Jiulongwan (McFadden et al., 2008)

Figure 3-5 The contact relationship between the Dengying Formation and the Doushantuo Formation at Huangniuya

3.4 Teaching Process and Precautions

3.4.1 Teaching Process

(1) Teachers should remind students to preview the Neoproterozoic stratigraphic sequence in South China in advance. At the starting point of this section, teachers should present a 15-minutes' introduction concerning the regional background and research status.

(2) Half of a day is allocated to route observation. The observation points are located at Shibanqiao (Gaojiaxi), Jiulongwan, and Huangniuya.

(3) This route represents the typical strata of the pre- and post-Neoproterozoic snowball Earth in South China (even the global), environmental events (oxygenic event, methane seeps event, marine acidification event), early life evolution events on Earth (the Doushantuo biota, e.g., Lantian, Weng'an, Miaohe, and Ediacaran biota), and mineralization events (phosphorite, shale gas, ect.). In the process of teaching, instructors explain the geological background, global correlation, and scientific significance of exposed strata in this route.

(4) Every student is required to sketch the crude stratigraphic column while observing the outcrop. The draft should be finished before the end of internship.

3.4.2 Precautions

(1) Every teacher and student should make safety a priority since the route is a winding mountain road.

(2) The vehicle will follow through, so the driver must be experienced and specialized in dealing with winding roads.

(3) Bring enough drinking water.

3.5 Focused Study and Reflections

(1) Recognize and analyze the origin and geological record of the methane seeps and its effect on paleo-ocean, paleoenvironment, and paleoecology after the snowball Earth event.

(2) Compare the stratigraphic correlation between the Liantuo Formation and the global Cryogenian strata.

(3) Analyze the origin, sedimentary process, provenance, and paleogeography significances of the tillite in the Nantuo Formation.

(4) Discuss about the origin, provenance, and process of phosphorus mineralization during Doushantuoian.

(5) Explore the environmental conditions and life evolution of the Doushantuo biota, the Ediacaran biota, and the Cambrian explosion.

4 ROUTE TWO: OBSERVING THE SINIAN–CAMBRIAN STRATIGRAPHY AND PALEONTOLOGY

4.1 Teaching Route

Zigui Base–Zhoujia'ao–Gunshi'ao–Zhoujia'ao–Zigui Base

4.2 Teaching Tasks and Requirements

(1) Understanding the lithological differences and boundaries between the Shibantan Member and the Baimatuo Member of the Sinian Dengying Formation.

(2) Observing the lithological differences and boundaries of the Sinian Dengying Formation and the Cambrian Yanjiahe Formation.

(3) Observing the stratigraphic sequences of the Yanjiahe Formation and the Shuijingtuo Formation (Fortunian and Age 2, Cambrian), and learning features of disconformity in the field and then drawing sketches.

(4) Studying fossil collection.

4.3 Route Information and Observing Points

The route includes three sections: the Shibantan Member and the Baimatuo Member of the

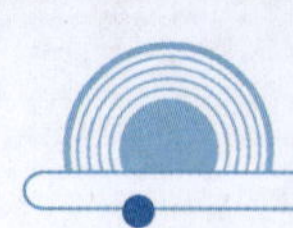

Sinian Dengying Formation, the Cambrian Terreneuvian Yanjiahe Formation, and the Series 2 Shuijingtuo Formation (Figure 4-1).

4.3.1 The Dengying Formation (Z_2dy)

In Route One, the stratigraphic characteristics of the Hamajing Member and the Shibantan Member of the Dengying Formation have already been examined. Route Two focuses on the boundary characteristics of the Shibantan Member and the Baimatuo Member, and the lithostratigraphic characteristics of the Baimatuo Member.

The observation point is located at the turn of the Tusan road, about 1 km southeast of Zhoujia'ao, and about 200 m north of the dolomite quarry (Figure 4-1). The Shibantan Member (Z_2dy^s) is featured by greyish-black medium–thin-bedded limestone alternated with greyish-black silty limestone and dolomitic limestone in unequal thicknesses.

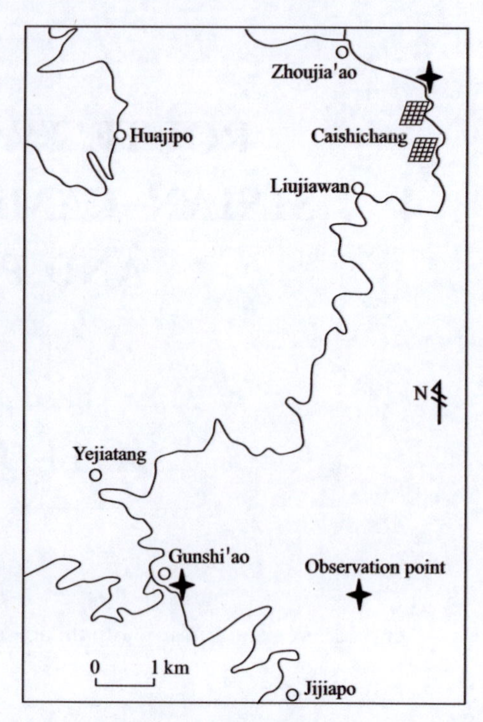

Figure 4-1 The traffic location map of observing points

The Baimatuo Member (Z_2dy^b) is marked by a large amount of grayish-white thin-bedded dolomites, with a thickness of 75.67–469 m. The lower part is composed of greyish-white thick–medium-bedded dolomite, calcite dolomite, gravel–bearing dolomite, and siliceous dolomite alternated with dolomitic limestone, containing siliceous nodules. The lower middle part consists of greyish-white–yellow medium-bedded dolomite alternated with siliceous dolomite and siliceous rocks, with siliceous nodules and bedded chert. The upper middle part is composed of pink–greyish-white medium thick-bedded doloarenite, containing siliceous nodules and bedded chert, and tabular cross-bedding and bird's-eye structure. The upper part is greyish-white thick-bedded dolomite, alternated with thin–medium-bedded silty dolomite, containing siliceous bands, siliceous nodules, and dolomitic nodules. This member contains a few acritarchs: *Asperatopsophosphaera partialis*, *Trachysphaeridium hyalinum*, *T. rude*, *Taeniatum crassum*, etc. Generally, it is deposited in tidal flat environment and conformably overlays the Shibantan Member.

4.3.2 The Yanjiahe Formation ($Є_1y$)

The Yanjiahe Formation ($Є_1y$) is mainly distributed in the Wuhe, Yanjiahe, Sixi, etc. The lower part is grey argillaceous dolomite, dolomite alternated with yellow carbonaceous mudstone with intercalation of 4–10 cm greyish-black siliceous bands, among which dolomite contains small shelly fossils. The upper part is greyish-black medium–thin-bedded limestone and carbonaceous shale intercalating carbonaceous limestone. The medium–thin-bedded carbonaceous limestone contains siliceous-phosphatic nodules whose diameters range from 5 cm to 8 cm, and the top part is grey medium-bedded limestone, containing siliceous nodules on which lays the 5–10 cm adobe claypan. This formation is about 56 m in thickness and is in disconformity with the underlying Dengying Formation and the overlying Shuijingtuo Formation (Figures 4-2 and 4-3).

Time	Lithostratigraphy		Thickness/m	Lithological Log (1/500)	Lithological Description
	Formation	Bed			
Series 2	the Shuijingtuo Formation				greyish-black carbonaceous siltstone and silty mudstone, alternated with dolomitic limestone lens. The lower part is composed of 2–3 cm grey and yellowish-brown claystone (weathering crust)
Terreneuvian	the Yanjiahe Formation	5	8		greyish-black medium–thin-bedded carbonaceous micrite–powder-crystalline limestone, alternated with silty mudstone, siliceous-phosphatic nodules in limestone. The top part is black thin-bedded siliceous rock, including small shelly fossils, trace fossils, etc. Small shelly fossils belong to the *Lophotheca-Aldanella-Maidipingoconus* assemblage
		4	17		darkish-grey medium–thin-bedded carbonaceous powder limestone, alternated with black thin-bedded carbonaceous and silty mudstone, including small shelly fossils of *Circotheca-Anabarites-Protohertzina*
		3	12		grey–dark grey thin-bedded silty siltstone, alternated with yellowish-green thin-bedded silty mudstone, with chert banding (the upper middle part is composed of grey medium-bedded dolomite)
		2	10		grey–dark grey medium-bedded siliceous dolomite, alternated with yellowish-green silty mudstone (the top part is greyish-black banded cherts)
		1	9		grey thin-bedded dolomite alternated with yellowish-green mudstone; the upper part is grey thin-bedded silty siltstone and silty mudstone, with chert banding
the Ediacaran Dengying Formation					grey thick–medium-bedded crystalline powder dolomite

Figure 4-2 Stratigraphic column of the Yanjiahe Formation

Figure 4-3 The Sinian Dengying Formation and the Cambrian Yanjiahe Formation in Gunshi'ao

A. the stratigraphic contact relationship betweenthe Dengying Formation and the Yanjiahe Formation;
B. the stratigraphic features of the lower Yanjiahe Formation

The main fieldwork tasks include:

(1) Observing the lithological and sedimentary characteristics of the top of the Dengying Formation.

(2) Observing the stratigraphic contact relationship between the Dengying Formation and the Yanjiahe Formation (Figure 4-3), i. e., stratigraphic features, lithological features, and depositional features.

(3) Observing the fault structure near the boundary between the Dengying Formation and the Yanjiahe Formation (Figure 4-3).

(4) Observing the stratigraphic sequence, its lithological features, and depositional features of the Yanjiahe Formation.

(5) Observing the siliceous nodule, lithological features, and features of small shelly fossils at the top of the Yanjiahe Formation.

4.3.3 The Shuijingtuo Formation ($\epsilon_2 s$)

The lowest boundary of the Shuijingtuo Formation is characterized by the appearance of blackish-grey thin-bedded carbonaceous siltstone and silty mudstone. The stratigraphy of the internship base ranges from 53 m to 161 m. The lower part is composed of black thin-bedded carbonaceous shale and silt mudstone, alternated with siliceous dolomite, and dolomite, dolomitic limestone lens; the middle part is blackish-grey and geryish-yellow carbonaceous shale and silty shale, alternated with thin–medium-bedded limestone; the upper part is blackish-grey thin–medium-bedded limestone with intercalations of thin-bedded muddy limestone and calcareous shale; the top is grey thin-bedded dolomitic limestone containing phosphatic nodules and calcareous dolomite, horizontal bedding and trilobites produced: *Tsunyidiscus ziguiensis*, *T. xiadongensis*, *Hupeidiscus orientalis*, *H. fengdongensis*, *H. latus*, etc. This formation is deposited unconformably on the underlying Yanjiahe Formation (Figures 4-4 and 4-5).

Time	Lithostratigraphy		Thickness/ m	Lithological Log (1/500)	Lithological Description	Sedimentary Facies
	Formation	Bed				
Series 2	the Shipai Formation				greyish-brown thin-bedded silty siltstone alternated with silty mudstone	
	the Shuijingtuo Formation	9	86		grey medium–thin-bedded micrite–powder-crystalline limestone, alternated with yellowish-green extremely thin mudstone, horizontal bedding in limestone, cross-bedding of small scale on the top	shallow shelf
		8	10		dark grey thin-bedded micrite–powder-crystalline limestone, alternated with greyish-black thin-bedded silty mudstone, including silty mudstone, horizontal bedding, *Tsunyidiscus* sp. produced	
		7	11		greyish-black thin-bedded micrite–microcrystalline limestone alternated with greyish-black carbonaceous and silty mudstone, horizontal bedding, with brachiopod and trilobite fossils occasionally seen	deep shelf
		6	18		greyish-black thin-bedded carbonaceous, calcareous, and silty mudstone, horizontal bedding, many fossils (such as trilobites and sponge spicules) produced	
		5	3		black thin-bedded carbonaceous and calcareous mudstone, siliceous and carbonaceous micritic limestone, and carbonaceous and silty mudstone	
		4	10		black thin-bedded carbonaceous and silty mudstone, horizontal bedding, including trilobite fossils	
		3	4		black medium–thin-bedded siliceous mudstone and siliceous dolomite, alternated with thin-bedded silty and carbonaceous mudstone	
		2	15		black thin-bedded carbonaceous mudstone, horizontal bedding, alternated with dolomitic limestone lens	deep shelf
		1	2		the lower part is composed of grey and yellowish-brown claystone, the rest is greyish-black silty and carbonaceous mudstone	subtidal zone
Terreneuvian	the Yanjiahe Formation				greyish-black medium–thin-bedded carbonaceous micriic limestone, including siliceous-phosphatic nodules or black vety thin-bedded siliceous rock, small shelly fossils can be seen	

Figure 4-4 Stratigraphic column of the Shuijingtuo Formation (Series 2, Cambrian)

The main fieldwork tasks includes:

(1) The stratigraphic contact relationship between the Shuijingtuo Formation and the Yanjiahe Formation (Figure 4-5), i. e., stratigraphic features, lithological features, and depositional features.

(2) The stratigraphic sequence, lithological features, and depositional features of the Shuijingtuo Formation.

(3)Collecting fossils such as sponge spicules, sphenothallus, and trilobites in the black carbonaceous mudstone of the Shuijingtuo Formation (Figure 4-5).

Figure 4-5 The Yanjiahe Formation and the Shuijingtuo Formation (Cambrian)

A, B. the stratigraphic contact relationship between the Yanjiahe Formation and the Shuijingtuo Formation at Gunshi'ao; C. pyritized sponge spicule of the Shuijingtuo Formation; D. the stratigraphic contact relations between the Yanjiahe Formation and the Shuijingtuo Formation at Jiuqunao

4.4 Teaching Process and Precautions

4.4.1 Teaching Process

(1) Teachers remind the students to preview the geological history and the Cambrian stratigraphic sequence in South China one day in advance.

(2) Teachers briefly describe the tasks, objectives, and requirements at the starting point of the section.

4.4.2 Precautions

Bring a magnifier and some fossil wrapping paper.

4.5 Focused Study and Reflections

(1) How is "the Cambrian explosion" shown in the Shuijingtuo Formation in the study area?

(2) What are the geological characteristics of shale gas of the Shuijingtuo Formation?

(3) How about the characteristics of the Ediacaran microbiota and their sedimentary environment?

5 ROUTE THREE: OBSERVING THE ORDOVICIAN DAPINGIAN STRATA AT HUANGHUACHANG, YICHANG

5.1 Teaching Route

Zigui Base–Alongside Yixing Road, Huanghuaxiang–Zigui Base

5.2 Teaching Tasks and Requirements

(1) Understanding the lithostratigraphic and cyclostratigraphic features in the Dawan Formation in the third period of Ordovician (Dapingian) and Early–Middle Ordovician, and gaining some preliminarily knowledge on how to identify the various types of limestones.

(2) Observing and discerning different types of fossils, such as calathids, oncocerida, and ammonoids, and drawing field sketches.

(3) Describing the lithologic features and paleontological compositions of carbonatites according to their strata, and sketching stratigraphic sections and columns of the Honghuayuan and Dawan Formations.

5.3 Route Information and Observing Points

5.3.1 The Honghuayuan Formation (O_1h)

The Honghuayuan Fromation was formed during Early Ordovician. The formation on this route is mainly composed of dark gray or grayish-yellow medium–thick-bedded bioclast limestones. In the formation, each layer consists one or more small cyclostratigraphies (parasequences), whose main component is medium–thick-bedded bioclastic limestones. Besides, all small cyclostratigraphies are shallowing-upward progradational parasequences (Figure 5-1). The lower part of the formation contains conodonts of the *Archaeoscyphia, Calathium* and *Serratognathus diversus* Biozone, while the upper part contains conodonts of the *Oepikodus communis* biozone and chitinozoas of the *Lagenochitina esthonica* chitinozoan Biozone, which extends all the way to the lower part of the overlying Dawan Formation. Besides, calathids and crinoids can also be seen at the section (Figure 5-2).

Figure 5-1 The lithologic features of the Honghuayuan Formation at the Huanghuachang section, Yichang

A. medium–thick-bedded small cyclostratigraphics within the 6th and 7th layers; B. grayish-yellow medium–thick-bedded cyclostratigraphics in the 8th layer, containing abundant crinoids and calathids

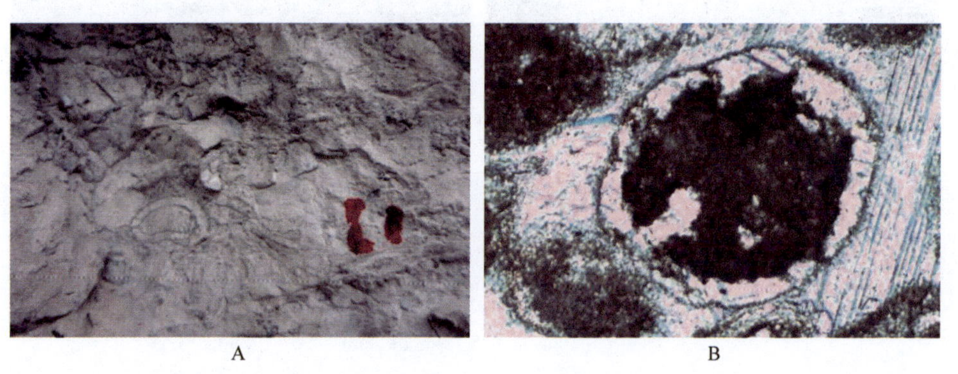

Figure 5-2 Fossils in the Honghuayuan Formation at the Huanghuachang section, Yichang

A. calathids, brachiopods, and others in the 2nd layer; B. calathids in the thin section of rock (minimum scale at the lower right corner: 20μm)

The Honghuayuan Formation contains abundant fossils. It is a key stratum to observe fossils for field trips. The main fieldwork tasks includes:

(1) Observing crinoid stems-contained clastic limestones, including its monocrystalline features, shapes, sizes, and contents.

(2) Observing calathid fossils, including its shapes, textures, sizes, and contents.

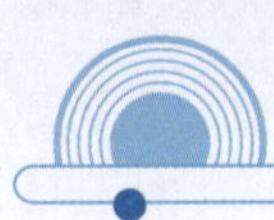

5.3.2 The Dawan Formation (O_{1-2d})

The Dawan Formation was formed during Early–Middle Ordovician. The section of the Dawan Formation can be divided into four members according to its lithologic features.

The First Member, with a thickness up to 4.84 m, is composed of glauconite-bearing bioclastic limestones associated with yellowish-green and thin-bedded silty shales. In this member, we can see multiple cyclostratigraphies (parasequences), which are made of thin-bedded silty shales and thin–medium-bedded glauconites-contained bioclastic limestones, and belong to shallowing-upward progradational parasequences (Figure 5-3). The Dawan Formation is characterized by the presence of both cold- and warm-water faunas and a mixture of graptolites, chitinozoas, acritarchs, conodonts, brachiopods, trilobites, and cephalopods (汪啸风, 1980; 汪啸风等, 1999, 2005). From bottom to top, the formation contains two conodonts biozones (the *Oepikodus evae* Zone and the *Periodon flabellum* Zone), graptolites of the *Didymograptellus bifidus* Biozone, and chitinozoas of the *Lagenochitina esthonica-Conochitina langei* Biozone.

Figure 5-3 The litholoic features of the First Member of Dawan Formation at the Huanghuachang section, Yichang

A. gray medium–thin-bedded glauconite-bearing bioclastic limestones in the 10th layer; B. multiple glauconite-bearing shallowing-upward progradational cyclostratigraphics in the 11th–13th layers, made up of gray thin–medium-bedded bioclastic limestones

The Second Member, with a thickness of 4.52 m, is composed of dark gray sandy shales, dark gray sandy mudstones, and gray thin-bedded bioclastic limestones. In this member, we can see multiple glauconite-bearing shallowing-upward progradational cyclostratigraphies (parasequences) (Figure 5-4). The top of the member is made of gray medium–thick-bedded bioclastic limestones. The member contains abundant brachiopods of *Leptella grandis*, graptolites of the *Azygograptus suecicus* Biozone (conodonts such as *Periodon flabellum*, *Baltoniotus protriangularis*, and *Drepanoistodus forceps*,) and two chitinozoa biozones (the *Conochitina langei* Biozone and the *C. pseudocarinata* Biozone).

The Third Member, with a thickness of 2.2 m, is composed of yellowish-green sandy shales and grayish-green thin-bedded nodular argillaceous bioclastic limestones. In this member, we can see multiple shallowing-upward progradational cyclostratigraphies (parasequences) (Figure 5-5). The member contains graptolites of the *Azygograptus suecicus* Biozone, trilobites such as *Pseudocalymene transversa*, and *Agerina elongate*, brachiopods of the *Euorthisin* Biozone, conodonts of the *Baltoniodus triangularis* Biozone, and chitinozoas of the *Belonechitina* cf. *henryi* Biozone.

Figure 5-4 The lithologic features of the Second Member of the Dawan Formation at the Huanghuachang section, Yichang

A. multiple shallowing-upward progradational cyclostratigraphics in the 22nd layer, composed of yellowish-green thin-bedded sandy shales and gray thin-bedded bioclastic limestones (From the First Member to the Second Member, the thickness of the limestone layer becomes thinner and the interbedded mudstone layers increase.); B. glauconites in the 25th layer of the rock thin section (minimum scale at the lower right corner: 50μm)

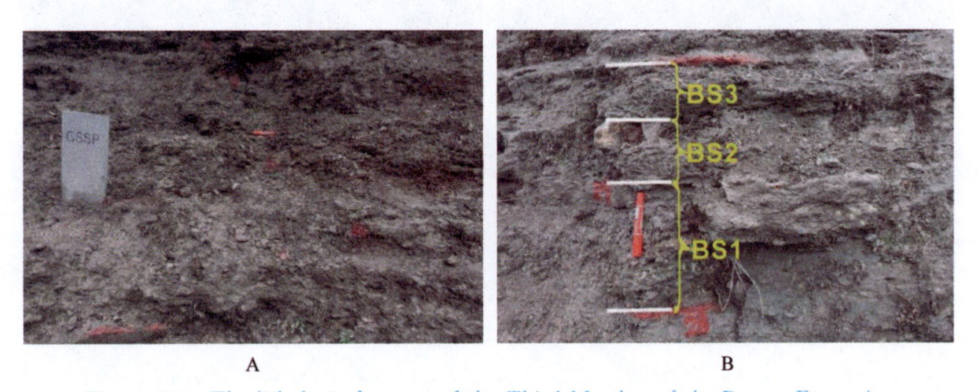

Figure 5-5 The lithologic features of the Third Member of the Dawan Formation at the Huanghuachang section, Yichang

A. the Dapingian GSSP between the 26th and 27th layers, with the 26th layer composed of yellowish-green sandy shales interbedded with grayish-green thin-bedded bioclastic limestones, and the 27th layer composed of grayish-green thin-bedded argillaceous bioclastic limestones interbedded with yellowish-green sandy shales; B. three small shallowing-upward progradational cyclostratigraphics in the 29th layer, composed of yellowish-green thin-bedded sandy shales and thin-bedded muddy bioclastic limestones

The Fourth Member, with a thickness of 9.52 m, is composed of yellowish green thin-bedded sandy shales, and grayish-red thin–medium-bedded muddy bioclastic limestones. In this member, we can see multiple shallowing-upward progradational cyclostratigraphies (parasequences) (Figure 5-6A and B). The member is abundant with glauconites. There are also graptolites of the *Azygograptus suecicus* Biozone, brachiopods of the *Euorthisina* biozone, conodonts of the *Baltoniodus navis* Biozone, and chitinozoas of the *Belonechitina* cf. *henryi* Biozone. On the whethered surface, megafossils such as armenoceras, ammonoids, and brachiopods can be seen (Figure 5-6C and D).

Figure 5-6 The lithologic features of the Fourth Member of the Dawan Formation at the Huanghuachang section, Yichang

A. a small cyclostratigraphy in the 31st layer, composed of yellowish-green–grayish-red medium-bedded muddy bioclastic limestones, containing glauconites; B. multiple shallowing-upward progradational cyclostratigraphics in the 39th layer, composed of grayish-red thin–medium-bedded muddy bioclastic limestones; C. abundant armenoceras, ammonoids, and brachiopods exposed at the weathered surface of the 39th layer; D. rachiopod shelln the the 39th layer of the rock thin section (minimum scale at the lower right corner: 20 μm)

The Dawan Fomation is well exposed. Compared with the underlying Honghuayuan Formation, the Dawan Formaton usually contains glauconites and many other fossils, and has obvious cyclostratigraphies (Figure 5-7). The major fieldwork tasks include:

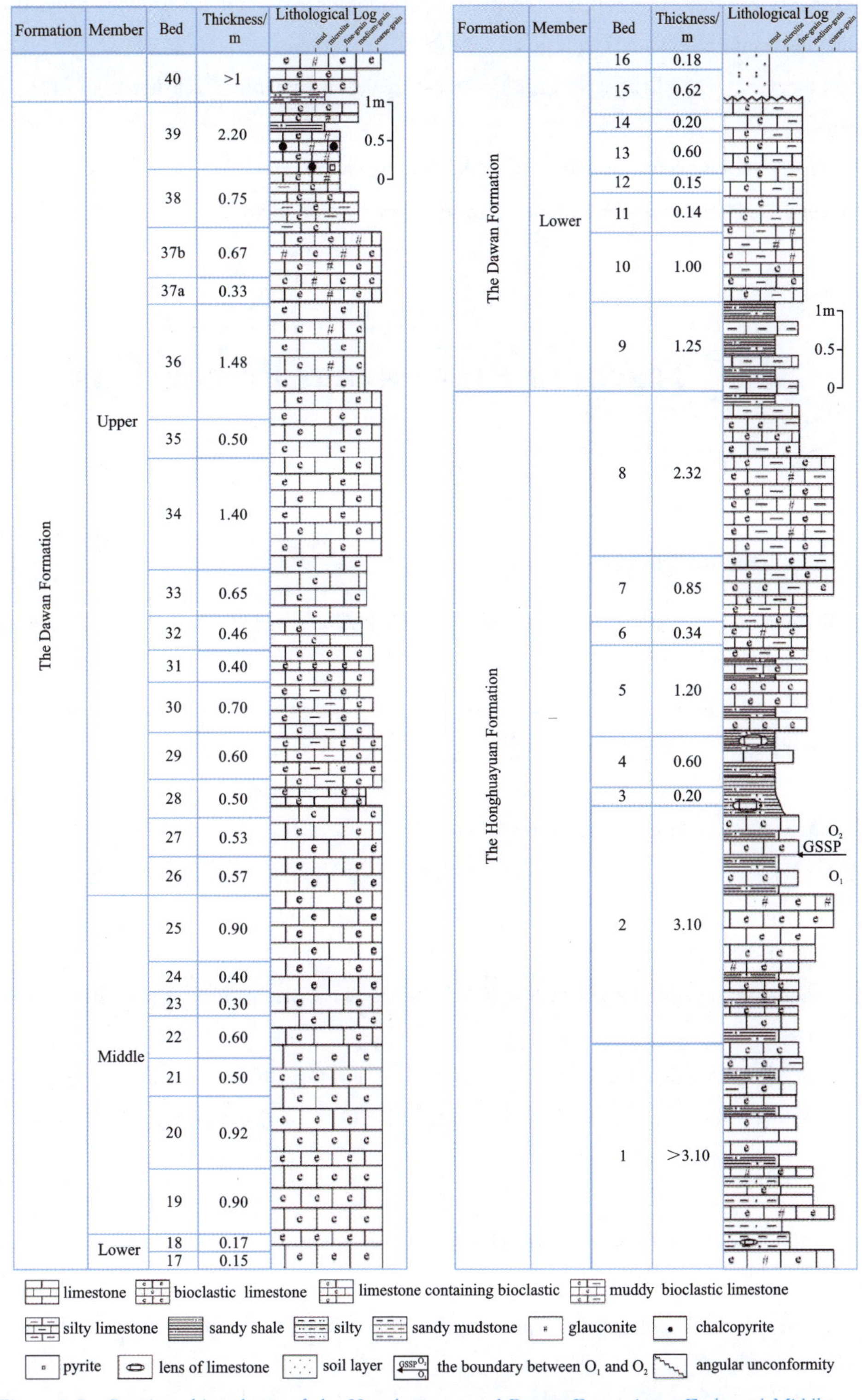

Figure 5-7 Stratigraphic column of the Honghuayuan and Dawan Formations (Early and Middle Ordovician) at the side of Yixing Road, Huanghuaxiang, Yichang, Hubei

(1) Observing and analyzing the contact relationships between the Dawan Formation and the underlying Honghuayuan Formation.

(2) Observing the lithostratigraphic and cyclostratigraphic features in the Dawan Formation.

(3) Observing bioclastic limestones, including the shapes, sizes, structures, and contents of armenoceras, ammonoids, brachiopods, and other fossils as well.

5.4 Teaching Process and Precautions

5.4.1 Teaching process

(1) Teacher reminds students to preview the Ordovician geological history and stratigraphic sequence of South China.

(2) Teacher briefs the tasks, objectives, and requirements of the field work at the starting point of the section. (5 minutes)

(3) Teacher introduces the Ordovician stratigraphic sequence of South China at the starting point of the section. (5 minutes)

(4) Teacher describes the stratification of the Honghuayuan Formation and sketches the section. (30 minutes)

(5) Teacher introduces Two observing points of the Honghuayuan Formation. (20 minutes)

(6) Teacher describes the stratification of the Dawan Formation and sketches the section. (60 minutes)

(7) Teacher introduces Four observing points of the Dawan Formation. (40 minutes)

(8) Teacher measures the Dapingjian "Golden Spikes" section. (60 minutes)

5.4.2 Precautions

(1) Students and teachers need to bring their own magnifiers. There are abundant fossils in the bioclastic limestones.

(2) Be careful while collecting samples and doing field observations, because the section is alongside the Yixing Road.

(3) Bring measuring line and tapes because it is required to measure the section on site.

(4) Bring food and drinking water.

5.5 Focused Study and Reflections

(1) Discuss about the forming environment of bioclastic limestones and sandy shales.

(2) Analyze the formation of glauconites.

(3) Explore the causes for the formation of nodular limestones.

6 ROUTE FOUR: OBSERVING THE UPPER ORDOVICIAN HIRNANTIAN GSSP IN WANGJIAWAN, YICHANG

6.1 Teaching Route

Zigui Base–Wangjiawan, Yichang–Zigui Base

6.2 Teaching Tasks and Requirements

(1) Understanding the stratigraphical sequence of the Upper Ordovician Wufeng Formation and the Lower Silurian Longmaxi Formation, and grasping the characteristics of black graptolite shale facies and shelly limestone facies.

(2) Learning the bottom boundary of the Hirnantian and the boundary between the Ordovician and the Silurian, and the relationships among lithostratigraphy, biostratigraphy, and chronostratigraphy.

(3) Describing the black shales in layers, and sketching the stratigraphic column of the Wufeng and the Longmaxi Formations.

(4) Mastering the methods of fossil collection and fossil identification.

6.3 Route Information and Observing Points

6.3.1 The Wufeng Formation (O_3w)

The Wufeng Formation was formed in the late Late Ordovician. In the Wangjiawan section, the Wufeng Formation is composed mainly of dark gray or grayish-black thin-bedded silicalites interbedded (in a cyclothem) with siliceous mudstones. The formation contains abundant graptolites.

The top of the Wufeng Formation is grayish-yellow medium-bedded argillaceous limestones, which contain abundant brachiopods and other shells. The lithology of the layer is special and can be found in Sichuan, Guizhou, Hubei, Shaanxi, and other provinces. However, the layer here is rather thin, with only 20 cm at the Wangjiawan section, which does not reach the thickness required to establish a formation. Therefore, the part is only called "the Guanyingqiao Bed".

The Guanyinqiao Bed (Formation) was named after the Guanyinqiao Bridge at Sichuan Province by Zhang Mingshao and Sheng Xinfu in 1939. It was later redefined by Lu Yanhao in 1959 (卢衍豪, 1959). The Guanyinqiao Bed (Formation) is widely distributed in Sichuan, Guizhou, Hubei, Shaanxi, and other provinces. As suggested, it is best to be called "the Guanyinqiao Bed", due to the fact that it is quite thin, and doesn't reach the thickness of a formation. The Guanyinqiao Bed at the Wangjiawan section, with a thickness of about 20 cm, is composed of grayish-yellow argillaceous limestones, containing benthic creatures such as brachiopods and trilobites (Figure 6-1). The lithology and biological compositions in the Guanyinqiao Bed are significantly different from those on the overlying and underlying strata. Moreover, the Guanyinqiao Bed is geographically widely distributed and of great significance for stratigraphic correlation.

The main fieldwork tasks include:

(1) Observing the silicalites and siliceous mudstones in the Wufeng Formation, including their colors, single bed thickness, sedimentary structures, and fossil types and preservation (distinguish the differences between in-situ and off-site burial).

(2) Observing the shell limestones in the Guanyinqiao Bed at the top of the Wufeng Formation, including their colors, single Bed thickness, and fossil types.

6.3.2 The Longmaxi Formation (O_3S_1l)

The Longmaxi Formation was mainly formed during the early stage of Early Silurian. The bottom, with a thickness of over 2 m, is grayish-black thin-bedded silicon-bearing carbonaceous mudstones, while the lower part is grayish-black carbonaceous shales with a thickness of about 50 m. At the teaching and observation points of the Wangjiawan section, only the bottom section of the Longmaxi Formation is exposed. The main observation points include the colors, single bed thickness, sedimentary structures, and fossil types of rocks in the Longmaxi Formation. Particular attention should be paid to the differences between the Wufeng and Longmaxi Formations with regard to their lithologic features and contact relationship.

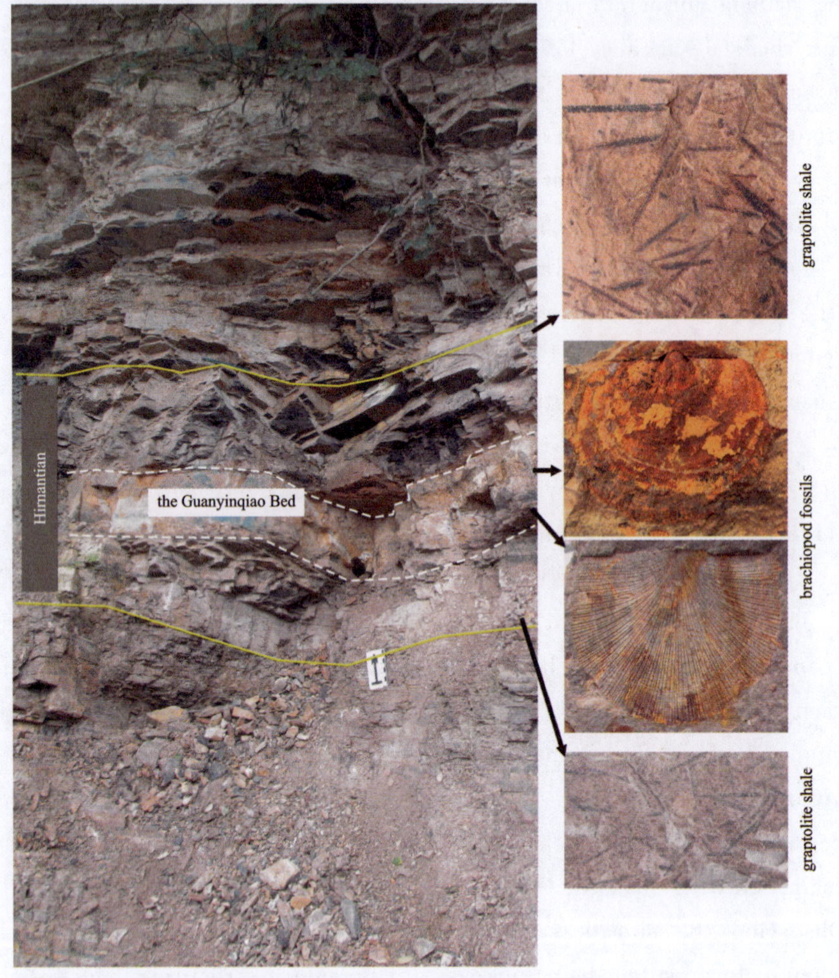

Figure 6-1 The lithostratigraphic texture and fossils of Hirnantian for the Wangjiawan section in Yichang
(provided by Wang Yongbiao)

6.4 Section Description and Fossil Collection

6.4.1 Section description

Section description includes mainly the description of colors, textures, features, sedimentary structures, and fossil types of rocks. Besides section description, stratigraphic columns are needed (To ensure the integrity and authenticity of the recording, it would be better to complete the stratigraphic columns in the field).

6.4.2 Fossil collection

Fossil collection in the field includes coding, packaging, and preliminary fossil identification. For further fossil identification after being back from the field, more references and documents might be consulted. According to the size of fossils, they could be divided into microfossils and macrofossils.

6.4.2.1 *Microfossil collection*

The collection of microfossils is comparatively easy: conodonts are usually preserved in limestones and marlstones, foraminifers and fusulinids in limestones, and radiolarians in silicalites or siliceous mudstones. Therefore, stratigraphic lithology is a key factor to be considered in microfossil collection. Besides, research purposes also need to be considered as regards to microfossil collection intervals. If the research purpose is to set up fossil zones on the basis of microfossils (e. g., conodonts), the possible stratigraphic thickness of each fossil zone needs to be calculated according to the division criteria and stratigraphic thickness of contemporaneous fossil zones in other regions or sections. It is required that multiple samples be collected at equal intervals in each fossil zone. And at the vicinity of possible chronostratigraphic boundaries, the sampling density needs to be increased. If the research purpose is to investigate paleoecology through microfossils, equidistant sampling needs to be adopted with reference to the thickness and lithological characteristics of the exposed strata at the section. For example, the strata of the Wufeng Formation contain radiolarians, but they are very thin, with a thickness of less than six meters. If conditions permit, samples can be collected in units of a single layer of silicalites or siliceous mudstones. The microfossil samples must be coded and registered (as shown in Table 6-1) upon collection. At the same time, it is necessary to fill the sample code in the corresponding places of the field book and measurement form. The coding of samples at field usually includes section name, stratigraphic position, and fossil type, HSR-2-C for instance. The code means it is a conodont (C) sample collected from the second layer (2) of strata at the section of Rencunping (R) in Sangzhi (S), Hunan Province (H).

6.4.2.2 *Macrofossil collection*

1. Macrofossil collection and sample coding

After creatures died, their fossils would usually distribute randomly along the layer if there was no strong hydrodynamic transformation and transportation. Therefore, a flat-mouth geological hammer is generally used to strip the macrofossils along the layer.

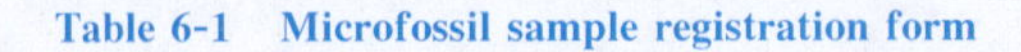

Table 6-1 Microfossil sample registration form

Section Name:

Section Code:

Coordinates of the Section Starting Point: E:　　　N:　　　H:

Collector's Working Place:

Collector:

Date Collected:　　　Year　　Month　　Day

Total Number of Collections:　　　Pieces

Sample Code	Sample Type	Lithologic Features	Location Collected		
			Layer No.	Distance from the bottom/m	Route (L): Collecting points/m

As soon as the fossil samples are collected, they should be numbered and registered in forms like Table 6-2. Sample codes should also be marked in corresponding places of the field book and measurement form.

Table 6-2 Macrofossil sample registration form

Section Name:

Section Code:

Coordinates of the Section Starting Point: E:　　　N:　　　H:

Collector's Working Place:

Collector:

Date Collected:　　　Year　　Month　　Day

Total Number of Collections:　　　Pieces

Sample Code	Field Naming	Preservation Condition	Surrounding-Rock Features	Location Collected		
				Layer No.	Distance from the bottom/m	Route (L): Collecting points/m

Coding macrofossil samples at field usually include section name, stratigraphic sequence, and fossil type, e.g., YWJW-W-2-B. The code means it is a brachiopod (B) sample collected from the second layer (2) of strata at the section of the Wufeng Formation (W) in Wangjiawan (WJW), Yichang (Y). Besides, it is also possible to add the year information in front of the code, for example 2014YWJW-W-2-B.

2. Category identification and description of macrofossils

In the field, it is necessary to at least identify the category of the macrofossil collected (such as trilobita, ammonoidea, and brachiopoda). If possible, it would be better to specify the family and genus, or even the species the sample belongs to. The description of the macrofossil may include fossil type, quantity (if measurable, otherwise it can also be described as "large" "medium" or "small" amount), and preservation condition (integrity, wearability, sortability, preservation status, etc.).

3. Macrofossil field photography

Macrofossil field photography refers to the photographing of the preservation status, ecotype, and key genera and species of the macrofossils.

To photograph the preservation status and ecotype of the fossils, it is better to take photos of multiple representative places, and at the same time, take good notes of their locations and sizes. It is also necessary to demonstrate the density and output features of fossils in the stratum.

To photograph the key genera and species of the fossils, the photos need to clearly reflect the overall shape, external structure, and texture of the fossils.

Besides, it is also necessary to choose a scale and an orientation according to the size of fossils in photographing.

4. Quantity control in macrofossils collection

In the field collection process, how do we know the number of fossils collected satisfies the needs of our research?

Rarefaction is one possible method to examine the point. However, the method needs to be carried out after one returns from the field and gets the fossil identification and statistical results in the lab. The approach is to enter the number of species and the number of samples of each species into "Past" software to produce a diagram. If the ending part of the parabola in the diagram tends to be flat, it means the number of samples is enough. If not, it means the number of samples is still not enough and more species might be found in that place. Fossil collection experiences and rarefaction using experiences indicate that if around 10 species are identified in one layer, 150 or so samples need to be collected. If the amount of species reaches 15, around 300 samples are to be collected.

6.4.2.3 *Indoor treatment, identification, and description of macrofossils*

1. Indoor treatment

Macrofossils: Macrofossils are often covered by surrounding rocks. Therefore, fossil preparation tools or steel needles are needed to move the covered part of the fossil from surrounding rocks. Before processing, it's better to observe and get familiar with the form of the fossil first (If the fossil is smaller than 5 mm, we can use a microscope so as to examine it carefully.). We may cut out the surrounding rocks from a place far away from the fossil in a far-to-near order.

Besides, the shell structure or inner structure of some fossils has to observe through making thin section. For example, to observe the structures of foraminifers and fusulinids, one has to slice the fossil till their initial chambers are exposed. In fact, the inner structures of most brachiopods have to be observed through thin sections.

2. The identification and description of macrofossils

The identification of macrofossils often includes the following procedures.

(1) Mastering the basic structure and description method of a certain category of fossils.

(2) Consulting references to understand the profiles of a certain category of fossils in the studied area or adjacent areas in the geohistorical periods.

(3) Reconstructing and categorizing fossils. The shapes of fossils are usually incomplete. By combining the features of multiple fossils, one may get to know the external characteristics and inner structures of a certain category of fossils. Besides, the parts of some fossils are often preserved in scattered matters, for example trilobites. In such conditions, one has to be very familiar with the head, breast, and tail characteristics of the trilobite species so as to combine the scattered fossils together. Built upon this previous work, one can draw sketches to reconstruct the characteristics of each type of fossils.

(4) Depicting fossils in order. For example, to describe brachiopods, one can start from describing its overall shape (such as its top view, side view, and front view), then move to the external structures, and last describe their inner structures. While describing, particular attention needs to be paid to their main identifying features, that is, the distinctive or unique features of a species in their genus.

If you are to introduce a new species, you should read all the literature related to a certain genus (both domestic and foreign literature included) to confirm that all known species of a certain genus are significantly different from the samples studied.

6.5 Teaching Process and Precautions

6.5.1 Teaching process

(1) Teacher reminds students one day ahead of schedule to preview Late Ordovician–Early Silurian geological history and stratigraphic sequence in South China, introduce students the format of drawing field lithological columns, and require them to get prepared accordingly.

(2) Teacher briefs at the starting point of the section the tasks, the objectives, and the requirements. (5 minutes)

(3) Teacher briefs at the starting point of the section the origin and meaning of the name "Hirnantian", the time when the Hirnantian GSSP was approved, and the significance of establishing the Hirnantian. (15 minutes)

(4) Teacher guides students to review at the starting point of the section of Late Ordovician–Early Silurian stratigraphic sequence in South China. (2 minutes)

(5) Teacher introduces stratification methods of black shales at the section. (8 minutes)

(6) Students identify and describe the strata from the Wufeng Formation to the bottom of the Longmaxi Formation, and draw corresponding lithological columns.

Group and task division: The whole class is divided into five or six groups with each group composed of five or six students. In a group, one student is to describe the stratum and draw its section profile, one or two students are to observe and study the lithologic features, one is to measure the thickness, and one or two students are to observe and study the sedimentary structures. It is better for all the members within one group to take turns to do different tasks while describing different strata. In this way, every student would have chances to learn and master the methods of observing and describing the lithology and sedimentary structures of strata, and drawing corresponding lithological columns in the field. The group leader needs to coordinate and record the division of labor in each group.

Requirements: Each group is to submit one complete record at the field which will be examined and graded on site. After returning from the field, each student is to complete and submit the records.

Time: About 150 minutes.

(7) Students collect fossils and take photos in the field. During this period, teacher

briefly introduces the methods of fossil collection and photographing. At the final phase of fossil collection, teacher introduces the methods of fossil identification and requires students to identify the fossils collected after they return from the field.

Steps (6) and (7) can be carried out simultaneously.

6.5.2 Precautions

(1) Tools needed in the field: Everyone needs to bring a field book, a pencil, an eraser, and a triangle or straight ruler; each group needs to prepare a marker, a steel measuring tape, and some fossil wrapping paper.

(2) Take care because the section is located alongside the road.

(3) Bring meals.

6.6 Focused Study and Reflections

(1) How to distinguish between the Wufeng Formation and the Longmaxi Formation?

(2) What are the relationships among the lithostratigraphy, biostratigraphy, and chronostratigraphy?

(3) What major geological events had occurred at the turn of Ordovician–Silurian? And what are the main reasons for the biological extinction?

(4) Where are the Hirnantia shelly faunas distributed in the world? What factors might be related to the diachroneity of the Hirnantia shelly faunas?

6.7 More Information about the Hirnantian GSSP

Hirnant is a small place in Bala, Wales, UK. The geologists named it after "Hirnantian", which refers to a stratigraphic unit in the uppermost Ordovician. The term "Hirnantian" was first proposed by Bancroft (1933). He used the term to refer to the brachiopods-bearing (*Hirnantia*) limestones in the uppermost Ordovician. Later, Bassett et al. (1966) and Ingham & Wright (1970) revised the term and redefined it as the limestone or mudstone strata at the uppermost Ordovician, which contain brachiopods (such as *Hirnantia*) and *Dalmanella* and trilobites (such as *Dalmanitina*).

The proposal to establish a global boundary stratotype section and point (GSSP) at the base of the Hirnantian was accepted by the Ordovician branch of the International Commission on Stratigraphy (ICS) in October 2004, and by ICS in February 2006 after some revision and improvement. In May 2006, the proposal was officially approved by the International Union of Geological Sciences (IUGS).

The Hirnantian is the seventh stage of the Ordovician, that is, the uppermost GSSP in the Ordovician. With a short duration of 1.8 million years, the establishment of the Hirnantian is of great significance. It has recorded events such as the second largest scale of mass extinction since Phanerozoic Eon (85% species extincted), the climatic cooling caused by the South Pole ice sheet expansion, and the global sea level declining (戎嘉余, 1984; Sheehan, 2001). The records provide a consistent chronostratigraphic standard for precise correlation research of pertinent strata around the globe and also for research on global bioevents and environmental events.

6.7.1 Reasons for the recognition of the North Wangjiawan section as the Hirnantian GSSP

The Hirnantian GSSP was set 0.39 m below the base of the Guanyinqiao Bed, North Wangjiawan section, Yichang, China. The section was recognized as the Hirnantian GSSP because of the following reasons (陈旭等, 2006a and b).

(1) Both the sedimentary and the biostratigraphic sequences are continuous in this section.

(2) The outcrops of strata in this section arc intact in this section.

(3) Graptolite and shell fossils are abundant and well-preserved in this section.

(4) The litho- and bio-facies in this section are stable and have extensive potential for comparison and contrast.

(5) The geological structure of the Wangjiawan section is simple, with no faults, folds, or other deformations.

(6) There are volcanogenic clay rocks near the boundary, suitable for isotope dating.

(7) The transportation is convenient for fieldwork.

(8) The biostratigraphy of this section is well researched, and extensive stratigraphic correlation research has also been carried out. In-depth carbon isotope studies have been carried out on nearby sections, which provides a good auxiliary mark for stratigraphic correlation.

6.7.2 Description of geological sections

The Wangjiawan section is located at Wangjiawan Village, 42 km north of Yichang, Hubei , China (N30°58′56″, E111°25′10″; Figure 6-2). From bottom to top, the Wangjiawan section outcrops the Wufeng Formation and the Longmaxi Formation (Figure 6-3). The two formations are in conformable contact.

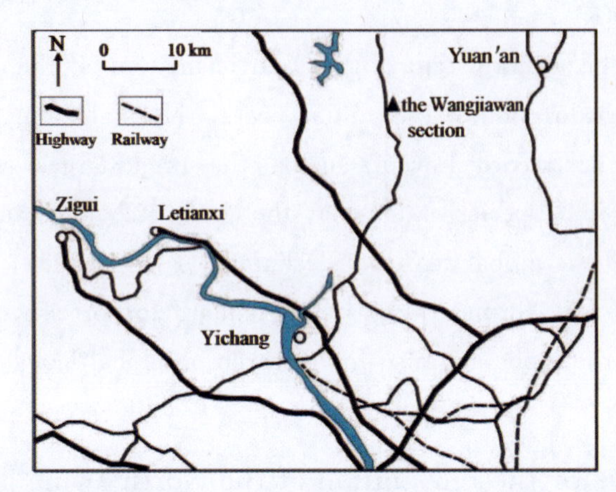

Figure 6-2 The location map of Wangjiawan section in Yichang

The Longmaxi Formation(O_3S_1l)

Bed 18 Grayish-black thin-bedded carbonaceous mudstones and silicon-bearing carbonaceous mudstones formed multiple cycles. The single layer thickness of the silicon-bearing carbonaceous mudstones increases from 3 cm to over 5 cm from bottom to top. This bed contains abundant graptolites. The bottom is not observed. **>50 cm**

Bed 17 Grayish-black thin-bedded carbonaceous mudstones and silicon-bearing carbonaceous mudstones formed into two cycles. The average thickness of a single layer of the silicon-bearing carbonaceous mudstones exceeds 4 cm. In the lower part, graded bedding is developed in the silicon-bearing carbonaceous mudstones. This bed contains abundant graptolites. **48 cm**

Bed 16 Grayish-black thin-bedded carbonaceous mudstones and silicon-bearing carbonaceous mudstones formed into two cycles. The average thickness of a single layer of the silicon-bearing carbonaceous mudstones exceeds 4 cm. In the carbonaceous mudstones, horizontal bedding is developed. This bed contains abundant graptolites. **17 cm**

Bed 15 Grayish-black thin-bedded silicon-bearing carbonaceous mudstones. The average thickness of a single layer exceeds 4 cm. In the lower part, horizontal bedding is developed. This bed contains abundant graptolites. **30 cm**

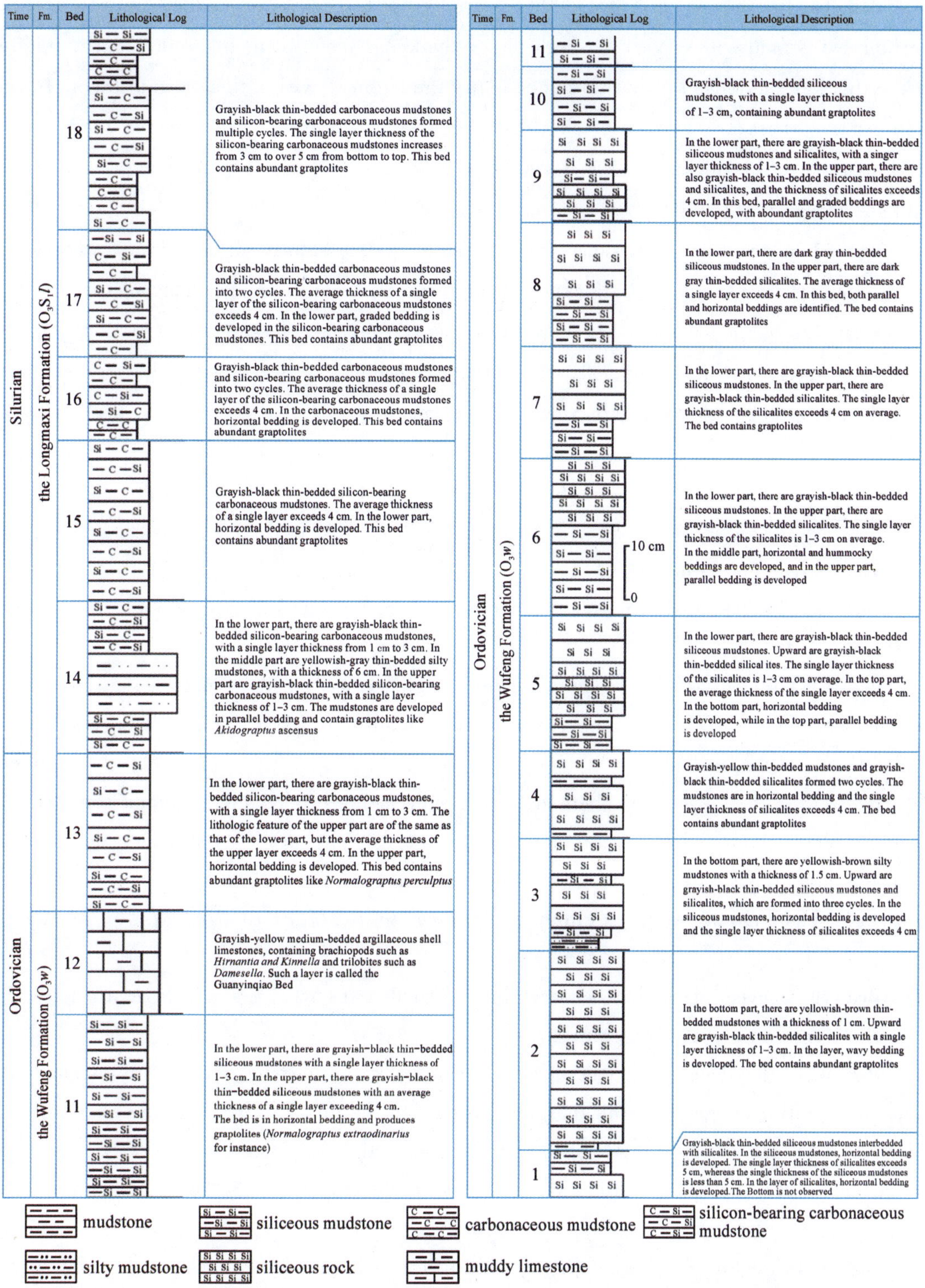

Time	Fm.	Bed	Lithological Log	Lithological Description
Silurian	the Longmaxi Formation (O_3S_1l)	18		Grayish-black thin-bedded carbonaceous mudstones and silicon-bearing carbonaceous mudstones formed multiple cycles. The single layer thickness of the silicon-bearing carbonaceous mudstones increases from 3 cm to over 5 cm from bottom to top. This bed contains abundant graptolites
		17		Grayish-black thin-bedded carbonaceous mudstones and silicon-bearing carbonaceous mudstones formed into two cycles. The average thickness of a single layer of the silicon-bearing carbonaceous mudstones exceeds 4 cm. In the lower part, graded bedding is developed in the silicon-bearing carbonaceous mudstones. This bed contains abundant graptolites
		16		Grayish-black thin-bedded carbonaceous mudstones and silicon-bearing carbonaceous mudstones formed into two cycles. The average thickness of a single layer of the silicon-bearing carbonaceous mudstones exceeds 4 cm. In the carbonaceous mudstones, horizontal bedding is developed. This bed contains abundant graptolites
		15		Grayish-black thin-bedded silicon-bearing carbonaceous mudstones. The average thickness of a single layer exceeds 4 cm. In the lower part, horizontal bedding is developed. This bed contains abundant graptolites
		14		In the lower part, there are grayish-black thin-bedded silicon-bearing carbonaceous mudstones, with a single layer thickness from 1 cm to 3 cm. In the middle part are yellowish-gray thin-bedded silty mudstones, with a thickness of 6 cm. In the upper part are grayish-black thin-bedded silicon-bearing carbonaceous mudstones, with a single layer thickness of 1–3 cm. The mudstones are developed in parallel bedding and contain graptolites like *Akidograptus* ascensus
Ordovician		13		In the lower part, there are grayish-black thin-bedded silicon-bearing carbonaceous mudstones, with a single layer thickness from 1 cm to 3 cm. The lithologic feature of the upper part are of the same as that of the lower part, but the average thickness of the upper layer exceeds 4 cm. In the upper part, horizontal bedding is developed. This bed contains abundant graptolites like *Normalograptus perculptus*
	the Wufeng Formation (O_3w)	12		Grayish-yellow medium-bedded argillaceous shell limestones, containing brachiopods such as *Hirnantia and Kinnella* and trilobites such as *Damesella*. Such a layer is called the Guanyinqiao Bed
		11		In the lower part, there are grayish-black thin-bedded siliceous mudstones with a single layer thickness of 1–3 cm. In the upper part, there are grayish-black thin-bedded siliceous mudstones with an average thickness of a single layer exceeding 4 cm. The bed is in horizontal bedding and produces graptolites (*Normalograptus extraodinarius* for instance)

Time	Fm.	Bed	Lithological Log	Lithological Description
Ordovician	the Wufeng Formation (O_3w)	11		
		10		Grayish-black thin-bedded siliceous mudstones, with a single layer thickness of 1–3 cm, containing abundant graptolites
		9		In the lower part, there are grayish-black thin-bedded siliceous mudstones and silicalites, with a singer layer thickness of 1–3 cm. In the upper part, there are also grayish-black thin-bedded siliceous mudstones and silicalites, and the thickness of silicalites exceeds 4 cm. In this bed, parallel and graded beddings are developed, with aboundant graptolites
		8		In the lower part, there are dark gray thin-bedded siliceous mudstones. In the upper part, there are dark gray thin-bedded silicalites. The average thickness of a single layer exceeds 4 cm. In this bed, both parallel and horizontal beddings are identified. The bed contains abundant graptolites
		7		In the lower part, there are grayish-black thin-bedded siliceous mudstones. In the upper part, there are grayish-black thin-bedded silicalites. The single layer thickness of the silicalites exceeds 4 cm on average. The bed contains graptolites
		6	10 cm / 0	In the lower part, there are grayish-black thin-bedded siliceous mudstones. In the upper part, there are grayish-black thin-bedded silicalites. The single layer thickness of the silicalites is 1–3 cm on average. In the middle part, horizontal and hummocky beddings are developed, and in the upper part, parallel bedding is developed
		5		In the lower part, there are grayish-black thin-bedded siliceous mudstones. Upward are grayish-black thin-bedded silical ites. The single layer thickness of the silicalites is 1–3 cm on average. In the top part, the average thickness of the single layer exceeds 4 cm. In the bottom part, horizontal bedding is developed, while in the top part, parallel bedding is developed
		4		Grayish-yellow thin-bedded mudstones and grayish-black thin-bedded silicalites formed two cycles. The mudstones are in horizontal bedding and the single layer thickness of silicalites exceeds 4 cm. The bed contains abundant graptolites
		3		In the bottom part, there are yellowish-brown silty mudstones with a thickness of 1.5 cm. Upward are grayish-black thin-bedded siliceous mudstones and silicalites, which are formed into three cycles. In the siliceous mudstones, horizontal bedding is developed and the single layer thickness of silicalites exceeds 4 cm
		2		In the bottom part, there are yellowish-brown thin-bedded mudstones with a thickness of 1 cm. Upward are grayish-black thin-bedded silicalites with a single layer thickness of 1–3 cm. In the layer, wavy bedding is developed. The bed contains abundant graptolites
		1		Grayish-black thin-bedded siliceous mudstones interbedded with silicalites. In the siliceous mudstones, horizontal bedding is developed. The single layer thickness of silicalites exceeds 5 cm, whereas the single thickness of the siliceous mudstones is less than 5 cm. In the layer of silicalites, horizontal bedding is developed. The Bottom is not observed

mudstone; siliceous mudstone; carbonaceous mudstone; silicon-bearing carbonaceous mudstone

silty mudstone; siliceous rock; muddy limestone

Figure 6-3 The measured stratigraphic column of Hirnantian GSSP, Wangjiawan, Yichang, Hubei

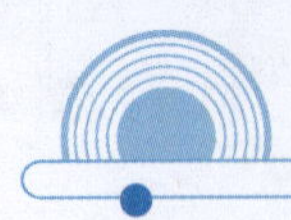

Bed 14 In the lower part, there are grayish-black thin-bedded silicon-bearing carbonaceous mudstones, with a single layer thickness from 1 cm to 3 cm. In the middle part, there are yellowish-gray thin-bedded silty mudstones, with a thickness of 6 cm. In the upper part, there are grayish-black thin-bedded silicon-bearing carbonaceous mudstones, with a single layer thickness of 1–3 cm. The mudstones are developed in parallel bedding and contain graptolites like *Akidograptus ascensus*. **29 cm**

Bed 13 In the lower part, there are grayish-black thin-bedded silicon-bearing carbonaceous mudstones, with a single layer thickness from 1 cm to 3 cm. The lithologic features of the upper part are of the same as that of the lower part, but the average thickness of the upper layer exceeds 4 cm. In the upper part, horizontal bedding is developed. This bed contains abundant graptolites like *Normalograptus perculptus*. **30 cm**

———— Conformable Contact ————

The Wufeng Formation (O_3w)

The Guanyinqiao Bed (In this handbook, the bed belongs to the top of the Wufeng Formation.)

Bed 12 Grayish-yellow medium-bedded argillaceous shell limestones, containing brachiopods such as *Hirnantia* and *Kinnella* and trilobites such as *Dalmesella*. Such a layer is called the Guanyinqiao Bed. **20 cm**

Bed 11 In the lower part, there are grayish-black thin-bedded siliceous mudstones with a single layer thickness of 1–3 cm. In the upper part, there are grayish-black thin-bedded siliceous mudstones with an average thickness of a single layer exceeding 4 cm. The bed is in horizontal bedding and produces graptolites (*Normalograptus extraodinarius* for instance). **39 cm**

Bed 10 Grayish-black thin-bedded siliceous mudstones with a single layer thickness of 1–3 cm, containing abundant graptolites. **12 cm**

Bed 9 In the lower part, there are grayish-black thin-bedded siliceous mudstones and silicalites, with a singer layer thickness of 1–3 cm. In the upper part, there are also grayish-black thin-bedded siliceous mudstones and silicalites, and the thickness of silicalites exceeds 4 cm. In this bed, parallel and graded beddings are developed, with aboundant graptolites. **17 cm**

Bed 8 In the lower partare dark gray thin-layered siliceous mudstones. In the upper part, there are dark gray thin-bedded silicalites. The average thickness of a single layer exceeds 4 cm. In this bed, both parallel and horizontal beddings are identified. The bed contains abundant graptolites. **23 cm**

Bed 7 In the lower part, there are grayish-black thin-bedded siliceous mudstones. In the upper part, there are grayish-black thin-bedded silicalites. The single layer thickness of the silicalites exceeds 4 cm on average. The bed contains graptolites. **20.5 cm**

Bed 6 In the lower part, there are grayish-black thin-bedded siliceous mudstones. In the upper part, there are grayish-black thin-bedded silicalites. The single layer thickness of the silicalites is 1–3 cm on average. In the middle part, horizontal and hummocky beddings are developed, and in the upper part, parallel bedding is developed. **28.5 cm**

Bed 5 In the lower part, there are grayish-black thin-bedded siliceous mudstones. Upward are grayish-black thin-bedded silicalites. The single layer thickness of the silicalites is 1–3 cm on average. In the top part, the average thickness of the single layer exceeds 4 cm. In the bottom part, horizontal bedding is developed, while in the top part, parallel bedding is developed. **25 cm**

Bed 4 Grayish-yellow thin-bedded mudstones and grayish-black thin-bedded silicalites form two cycles. The mudstones are in horizontal bedding and the single layer thickness of silicalites exceeds 4 cm. The bed contains abundant graptolites. **16 cm**

Bed 3 In the bottom part, there are yellowish-brown silty mudstones with a thickness of 1.5 cm. Upward are grayish-black thin-bedded siliceous mudstones and silicalites, which are formed into three cycles. In the siliceous mudstones, horizontal bedding is developed and the single layer thickness of silicalites exceeds 4 cm. **20.5 cm**

Bed 2 In the bottom part, there are yellowish-brown thin-bedded mudstones with a thickness of 1 cm. Upward are grayish-black thin-bedded silicalites with a single layer thickness of 1–3 cm. In the layer, wavy bedding is developed. The bed contains abundant graptolites. **37 cm**

Bed 1 Grayish-black thin-bedded siliceous mudstones interbedded with silicalites. In the siliceous mudstones, horizontal bedding is developed. The single layer thickness of silicalites exceeds 5 cm, whereas the single thickness of the siliceous mudstones is less than 5 cm. In the layer of silicalites, horizontal bedding is developed. (The bottom is not observed.) **>10 cm**

7 ROUTE FIVE: OBSERVING LATE ORDOVICIAN–PERMIAN STRATIGRAPHY AND PALEONTOLOGY

7.1 Teaching Route

Zigui Base–Wulong–Wenhua–Zigui Base

7.2 Teaching Objectives and Requirements

(1) Observing and describing Late Ordovician–Early and Mid-Silurian stratigraphic sequence, and the lithological characteristics of each formation.

(2) Observing and describing Devonian–Permian stratigraphic sequence, and the lithological characteristics of each formation.

(3) Drawing a stratigraphic column of the Upper Ordovician–the Permian.

7.3 Route Information and Observing Points

The stratigraphic sequences of the Wulong section are shown in Figures 7-1 and 7-2.

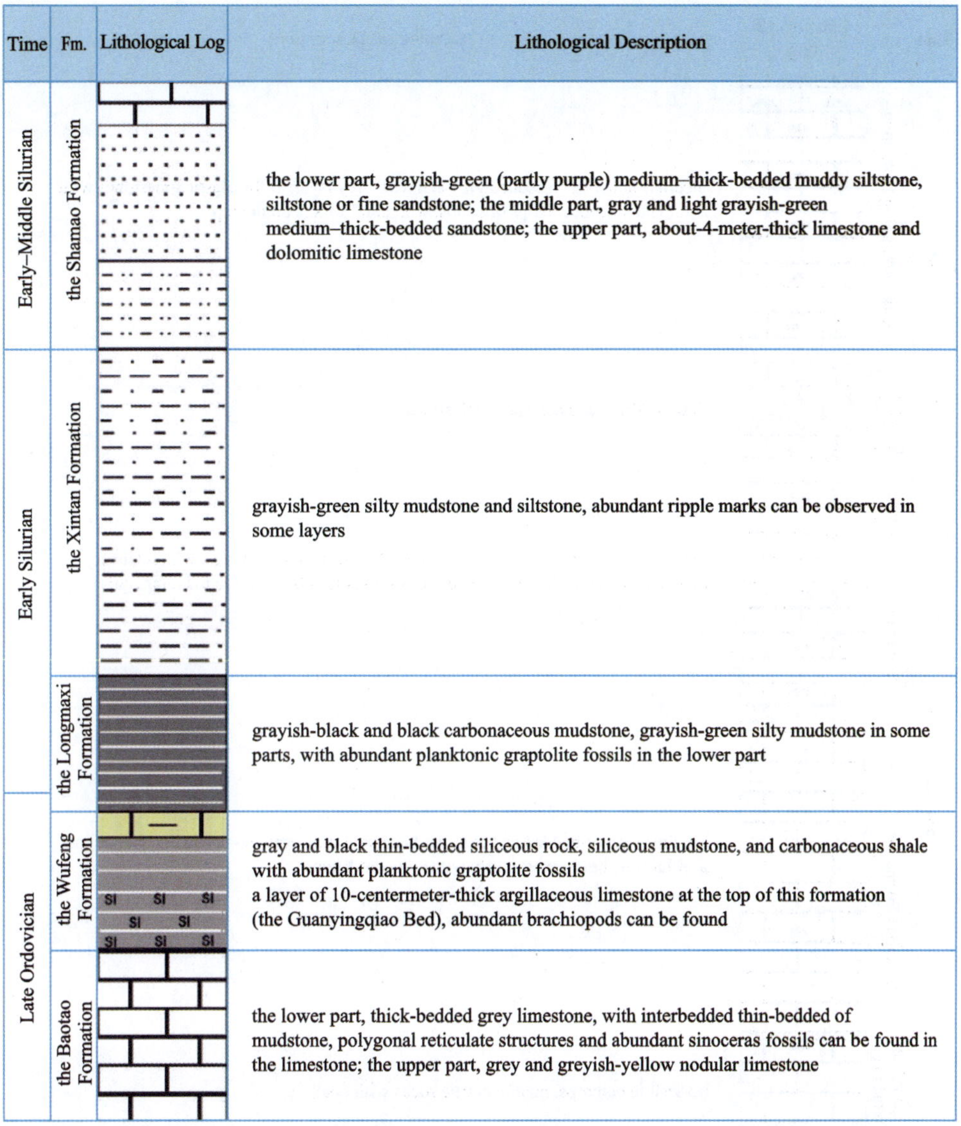

Time	Fm.	Lithological Log	Lithological Description
Early–Middle Silurian	the Shamao Formation		the lower part, grayish-green (partly purple) medium–thick-bedded muddy siltstone, siltstone or fine sandstone; the middle part, gray and light grayish-green medium–thick-bedded sandstone; the upper part, about-4-meter-thick limestone and dolomitic limestone
Early Silurian	the Xintan Formation		grayish-green silty mudstone and siltstone, abundant ripple marks can be observed in some layers
	the Longmaxi Formation		grayish-black and black carbonaceous mudstone, grayish-green silty mudstone in some parts, with abundant planktonic graptolite fossils in the lower part
Late Ordovician	the Wufeng Formation	SI SI SI SI SI SI SI SI	gray and black thin-bedded siliceous rock, siliceous mudstone, and carbonaceous shale with abundant planktonic graptolite fossils a layer of 10-centermeter-thick argillaceous limestone at the top of this formation (the Guanyingqiao Bed), abundant brachiopods can be found
	the Baotao Formation		the lower part, thick-bedded grey limestone, with interbedded thin-bedded of mudstone, polygonal reticulate structures and abundant sinoceras fossils can be found in the limestone; the upper part, grey and greyish-yellow nodular limestone

Figure 7-1 Stratigraphic column of the Wulong section from Late Ordovician to Early–Middle Silurian

7.3.1 The Baota Formation (O_3b)

The lower part of the Baota Formation is composed of grey thick-bedded limestones, with thin-bedded mudstone in between. Polygonal reticulate structures and abundant sinoceras fossils can be found in limestones. The upper part of the formation is composed of grey and greyish-yellow nodular limestones.

7.3.2 The Wufeng Formation (O_3w)

The Wufeng Formation is composed of grey and greyish-black thin-bedded siliceous

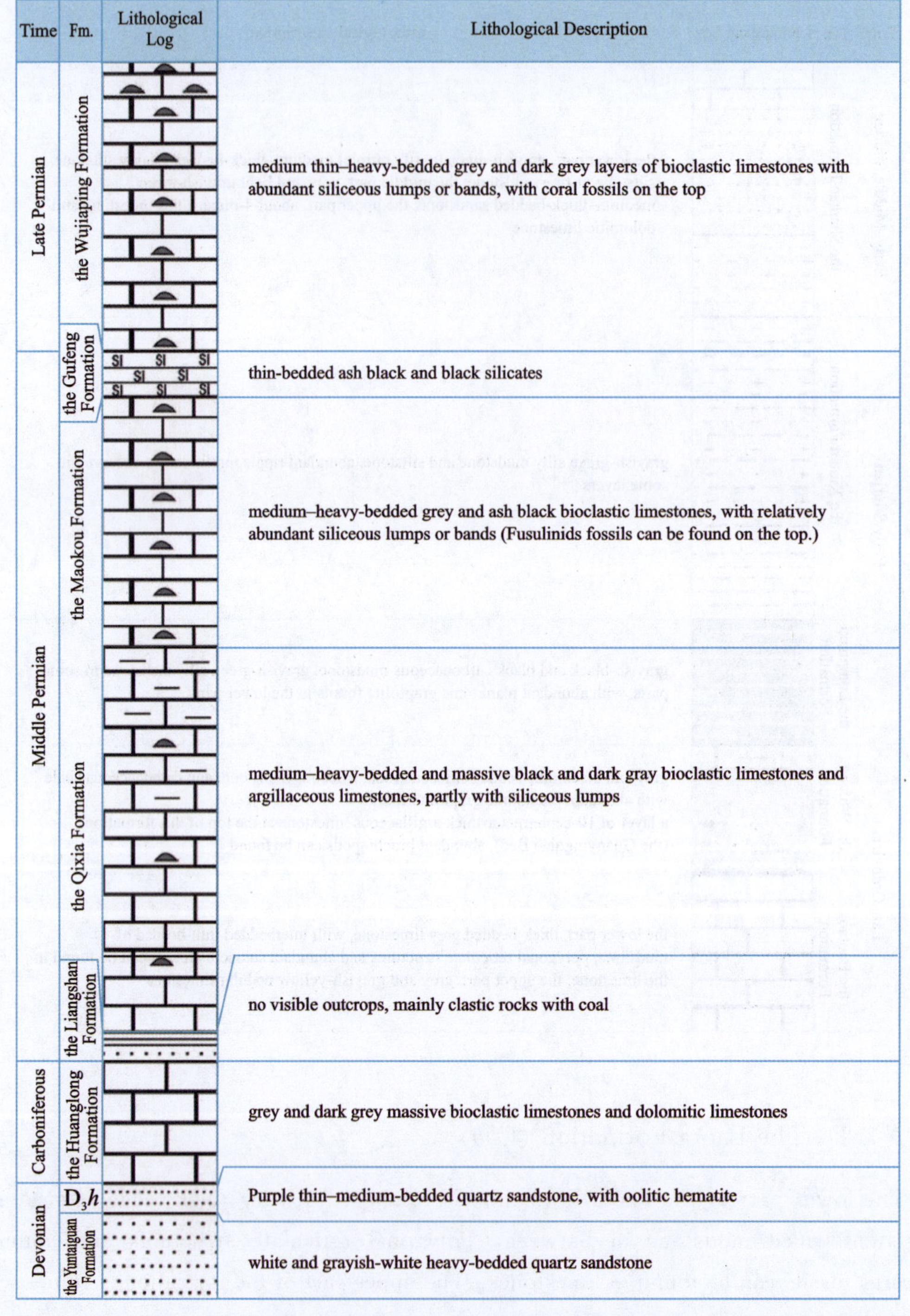

Figure 7-2 Stratigraphic column of the Wulong section from Late Paleozoic

rocks, siliceous mudstones, and carbonaceous shales, with abundant planktonic graptolite fossils. A 10 cm layer of greyish-yellow argillaceous limestones (the Guanyinqiao Bed) can be found on the upper part of the formation, with abundant benthic brachiopod fossils.

Reference: In the Wangjiawan section of Yichang, according to the zoning of graptolite fossils, about 30 cm above the top of the Guanyinqiao Bed is regarded as the boundary between the Ordovician and the Silurian.

7.3.3 The Longmaxi Formation ($S_1 l$)

The Longmaxi Formation is composed of greyish-black and black carbonaceous mudstones, partly greyish-green silty mudstones, with abundant planktonic graptolite fossils in the lower part of the formation.

7.3.4 The Xintan Formation ($S_1 x$)

The Xintan Formation is composed of greyish-green silty mudstones with ripple marks.

Reference: The Xintan Formation in this section is roughly equivalent to the Luoreping Formation in terms of occurrence, or it is beneath the Luoreping Formation. Unlike the Xintan Formation, the Luoraping Formation generally contains multiple layers of limestones.

7.3.5 The Shamao Formation ($S_1 sh$)

The lower part of the Shamao Formation is composed of greyish-green (partly purplish-red) medium–thick-bedded argillaceous siltstones, siltstones, and fine sandstones.

The middle part of the formation is composed of grey and light greyish-green medium–thick-bedded sandstones.

The upper part of the formation is composed of limestones and dolomitic limestones with a thickness of 4 m.

Reference: Between the Shamao Formation and its overlying Yuntaiguan Formation, there is a deposition gap from Middle and Late Silurian to Early Devonian. But the orientation in these formations are generally the same. Therefore, there is a disconformity between the underlying Shamao Formation and the overlying Yuantaiguan Formation.

7.3.6 The Yuntaiguan Formation ($D_2 y$)

The Yuntaiguan Formation is composed of white and greyish-white thick-bedded quartz sandstones.

7.3.7 The Huangjiadeng Formation ($D_3 h$)

The Huangjiadeng Formation is composed of purplish-red thin-bedded ferruginous quartz sandstones in which oolitic hematite can be found (Figure 7-3). Some parts are covered by Quaternary sediments in this formation.

Figure 7-3 Oolitic hematite-bearing quartz sandstones of the Devonian Huangjiadeng Formation

(The dark colored concentric grains are oolitic hematites. The white parts are quartz particles.)

7.3.8 The Huanglong Formation (C_2h)

The Huanglong Formation is composed of grey and dark grey massive bioclastic limestones and dolomitic limestones.

Reference: Between the Huanglong Formation and its underlying Yuntaiguan Formation, there is a deposition gap, Early Carboniferous deposits are missing. But the orientation of strata in these formations are generally the same. Therefore, there is a disconformity between the two formations.

7.3.9 The Liangshan Formation (P_2l)

No visible outcrops can be found in the Liangshan Formation. Based on an analysis of the slopewash from across the river and the data of nearby area, it can be concluded that the Liangshan Formation is mainly composed of clastic rocks with coal.

Reference: Based on regional geological setting, there might be a hiatus between the Liangshan Formation and its underlying Huanglong Formation. As the orientations of strata in these formations are generally the same, there is a disconformity between the two formations.

7.3.10 The Qixia Formation (P_2q)

The Qixia Formation is composed of black and dark gray medium–thick-bedded and massive bioclastic limestones and argillaceous limestones, partly with siliceous lumps.

7.3.11 The Maokou Formation (P_2m)

The Maokou Formation is composed of grey and ash black medium–thick-bedded bioclastic limestones, with relatively abundant siliceous lumps or bands. Fusulinids fossils can be found on the top.

7.3.12 The Gufeng Formation (P_2g)

The Gufeng Formation is composed of greyish-black and black thin-bedded silicates. Because of weathering, the area contains this formation usually has negative terrain.

7.3.13 The Wuchiaping Formation (P_3w)

The Wuchiaping Formation is compose of grey and dark grey medium thin–thick-bedded bioclastic limestones with abundant siliceous lumps or bands. Coral fossils can be found on the top.

7.3.14 The Daye Formation (T_1d)

The bottom of this formation is covered by Quaternary sediments. Based on regional geological data, it's estimated that the Daye Formation could be composed of mudstones or argillaceous limestones. The upper part of the formation is mainly composed of grey and greyish-white thin-bedded limestones, and fossils are rarely found in this formation.

Reference: Because of the end-Permian mass extinction, there are very few fossils in the Early Triassic strata.

7.4 Teaching Process and Precautions

(1) Students should preview the stratigraphic sequences of Ordovician, Silurian, Devonian, Carboniferous, Permian, and Early Triassic, as well as the lithologic characteristics of each formation.

(2) After finishing observing the stratigraphic sequences of Silurian, the teacher should

make a summary of Early Paleozoic stratigraphic sequences and then analyze the stratigraphic records of the Caledonian Movement.

(3) Remember to bring food and water.

7.5 Focused Study and Reflections

(1) What kinds of marine environment do the Guanyinqiao Bed and its overlying and underlying graptolite shales represent?

(2) How did the water depth change in Silurian indicated in carbonaceous mudstones in the Longmaxi Formation and sandstones in the Shamao Formation? What are the evidences?

(3) On top of the Maokou Formation, thin-bedded silicates of the Gufeng Formation occurred in a sudden. What kind of depositional environment change does it indicate?

(4) Why is that the boundary determined by lithostratigraphic units doesn't completely agree with the boundary determined by stratigraphic units and bio-stratigraphic units?

8 ROUTE SIX: OBSERVING PERMIAN STRATIGRAPHY AND PALEONTOLOGY IN LATE PALEOZOIC

8.1 Teaching Route

Zigui Base–Lvjiaping Tunnel Northwest Exit–Chain Cliff Village–Lvjiaping Tunnel Northwest Exit–Zigui Base

8.2 Teaching Tasks and Requirements

(1) Understanding the stratigraphic sequence of the Maokou Formation in Middle Permian and the Wuchiaping Formation in Late Permian; learning primarily the characteristics of different types of limestones.

(2) Observing the occurrence of siliceous agglomerates and drawing field sketches.

(3) Drawing a geological cross-section showing the disconformity betewwn the Middle and the Upper Permian.

8.3 Route Information and Observing Points

The stratigraphic sequence of the Maokou Formation (Middle Permian) to the Wuchiaping Formation (Late Permian) at Zigui, Hubei, China is shown in Figure 8-1.

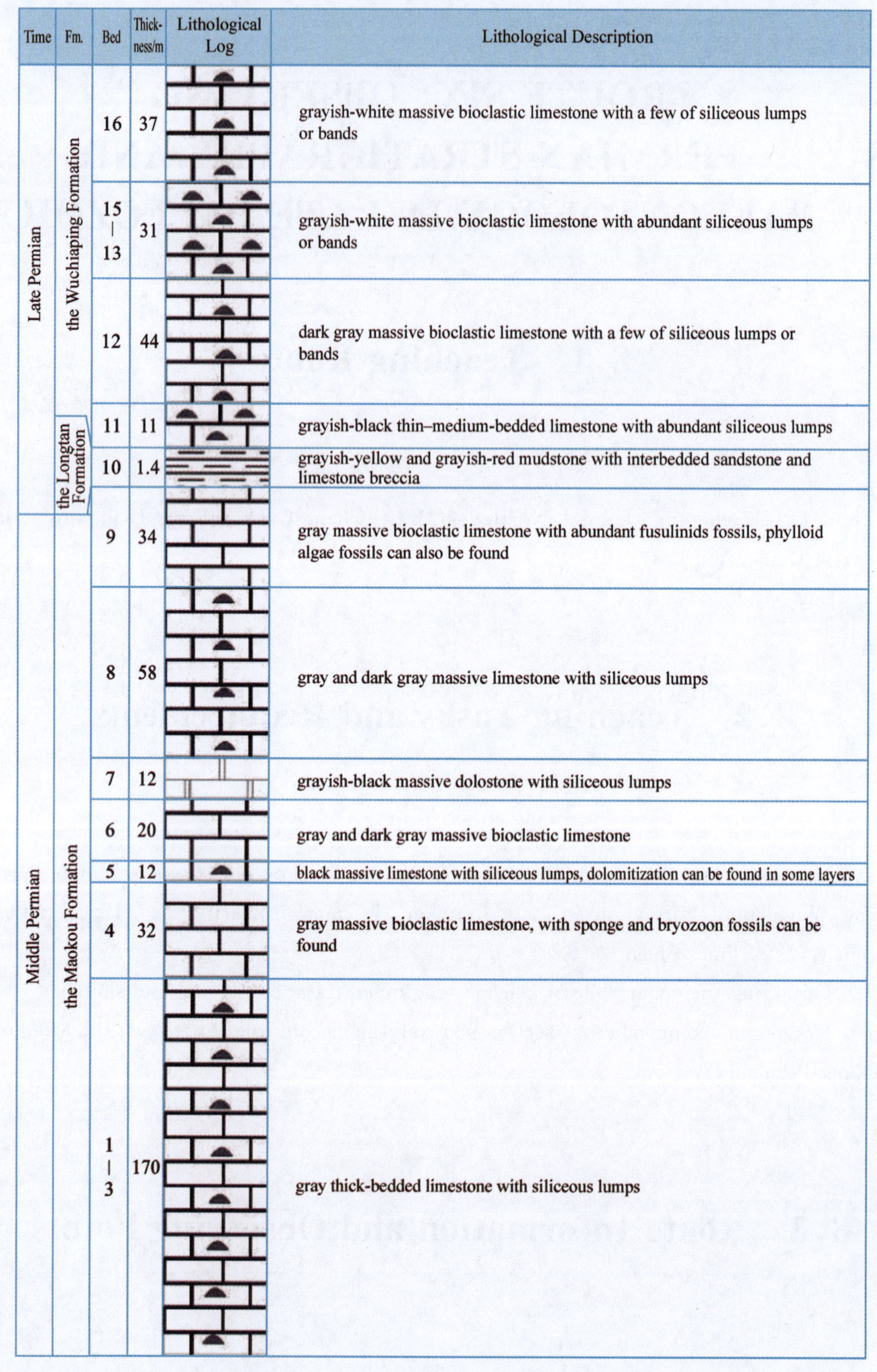

Time	Fm.	Bed	Thickness/m	Lithological Log	Lithological Description
Late Permian	the Wuchiaping Formation	16	37		grayish-white massive bioclastic limestone with a few of siliceous lumps or bands
		15–13	31		grayish-white massive bioclastic limestone with abundant siliceous lumps or bands
		12	44		dark gray massive bioclastic limestone with a few of siliceous lumps or bands
	the Longtan Formation	11	11		grayish-black thin–medium-bedded limestone with abundant siliceous lumps
		10	1.4		grayish-yellow and grayish-red mudstone with interbedded sandstone and limestone breccia
Middle Permian	the Maokou Formation	9	34		gray massive bioclastic limestone with abundant fusulinids fossils, phylloid algae fossils can also be found
		8	58		gray and dark gray massive limestone with siliceous lumps
		7	12		grayish-black massive dolostone with siliceous lumps
		6	20		gray and dark gray massive bioclastic limestone
		5	12		black massive limestone with siliceous lumps, dolomitization can be found in some layers
		4	32		gray massive bioclastic limestone, with sponge and bryozoon fossils can be found
		1–3	170		gray thick-bedded limestone with siliceous lumps

Figure 8-1 Stratigraphic column of the Middle Permian Maokou Formation–the Late Permian Wuchiaping Formation

8.3.1 The Maokou Formation (P_2m)

Roughly speaking, there are two different sedimentary facies during Late Permian in South China. The Maokou Formation is mainly a set of carbonate sediments rich in fossilized organisms or biological debris, representing the shallow-water carbonate platform environment. The Gufeng Formation is mainly a set of chert sediments which may contain radiolarians or sponge spicule fossils, representing relatively deep-water basin sediments. The Maukou Formation and the Gufeng Formation are different facies developed contemporaneously with the former as the upper part and the latter as the lower part in a section.

In this route, the lower part of the Maukou Formation is mainly composed of dark gray and grayish-black bioclastic limestone of medium thick-bedded–massive siliceous agglomerates, while the upper part is characterized by dark gray–light gray thick-bedded–massive bioclastic limestone , with fusulinid limestone and phylloid algae limestone. Under the influence of later diagenesis, dolomite or dolomitic clumps of limestone are developed in some parts of the strata.

The Maukou Formation is extremely rich in fossilized organisms, most of which require the aid of a microscope to identify. Those fossilized organisms can be observed in the outcrops, including different types of calcareous algae and foraminifera. The main fossils are crinoid stems, bryozoan, sponges, and fusulinid (Figure 8-2).

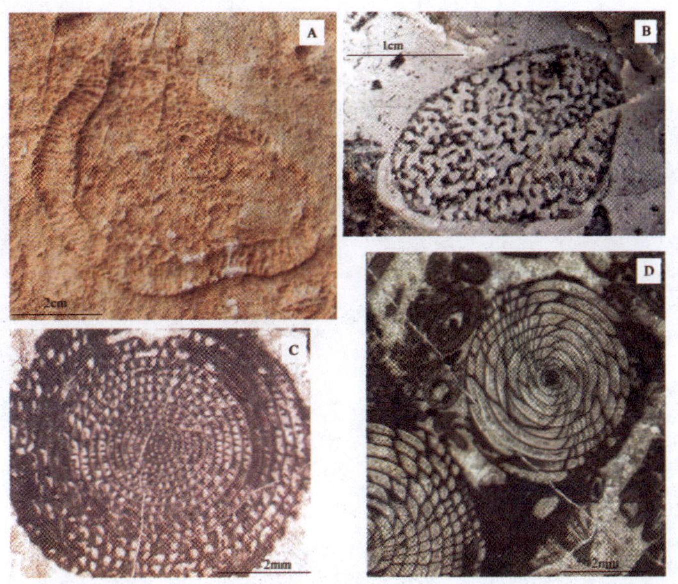

Figure 8-2 Major fossilized organisms of the Maokou Formation visible in the field

A, B. sponge fossils (field photo); C. *Neoschwagerina* (under the lens);

D. *Verbeekina* (under the lens)

The upper part of the Maukou Formation is the richest in fossils and is an ideal stratum for field fossil observation. The main fieldwork tasks include:

(1) Observing limestone with crinoid stems, i.e., the monocrystalline features, shapes, sizes, and contents.

(2) Observing sponge fossils, i.e., the shapes, structures, sizes, and contents.

(3) Observing dolomite clusters, i.e., the color, structures, and occurrence.

(4) Observing fusulinid limestone, i.e., the types, sizes, contents, colors, structures, and construction.

(5) Observing phylloid algae limestone, i.e., the shapes, structures, and preservation state.

8.3.2 The Longtan Formation (P_3l)

The typical Longtan Formation is a set of coal-bearing clastic sediments. On top of the grey rocks of the Maukou Formation, yellowish-brown and partly purplish-red sendiments of mudstone, sandstone, and limestone gravels stretches about 1.4 m in thickness. The overall sedimentary characteristics are similar to those of the Longtan Formation (Late Permian) in South China, which represents a set of clastic sediments after "the Dongwu Movement". It should be noted that the clastic section seen in this route lacks the thickness and does not contain coal system or coal seam, which differs from the typical coal-bearing clastic sediments of Longtan Formation. As a result, this section can also be classified as the clastic section of the lower Wuchiaping Formation. Due to the influence of "the Dongwu Movement", the unconformity between the Wuchiaping Formation and the underlying Maokou Formation is a disconformity (Figure 8-3).

Figure 8-3 The contact of the Maokou Formation and the Longtan Formation (field photo)

The main fieldwork tasks include:

(1) Observing and analyzing the contact relationship between the Maukou Formation and the Longtan Formation.

(2) Observing and describing the paleocrust of weathering.

8.3.3 The Wuchiaping Formation (P_3w)

The Wuchiaping Formation can represent both the early Late Permian or the entire Late Permian. The difference between the Wuchiaping Formation and the Longtan Formation is that the former is dominated by marine carbonate rocks, while the latter is dominated by coal-bearing clastic deposits in marine-terrestrial transitional facies. The difference between the Wuchiaping Formation and the Changxing Formation is that the former often contains abundant siliceous agglomerates, while the latter contains little or no siliceous agglomerates. The Wuchiaping Formation in this route is mainly composed of meidum-bedded siliceous agglomerated limestone, whose content in some layers is as high as 45% (Figure 8-4).

Figure 8-4 The siliceous agglomerated limestone of the Wuchiaping Formation (field photo)

Compared with the Maokou Formation on this route, although there are still abundant debris of fossilized organisms in the limestone of the Wuchiaping Formation, the large-scale fusulinid fossils have disappeared, and the sponge fossils have also been significantly reduced. However, more abundant brachiopod fossils can be seen in some layers.

The outcrop of this formation is well exposed, offering an ideal opportunity to observe the field characteristics of siliceous agglomerates and the contact relationship between the Maokou Formation and the Wuchiaping Formation.

The main fieldwork tasks include:

(1) Observing siliceous agglomerated limestone in the Wuchiaping Formation, i. e., describing the color, structure, shape, size, and content of the agglomerate, and drawing a field sketch.

(2) Observing the characteristics of limestone in the Wuchiaping Formation, i. e., describing the color, structures, and fossils while paying attention to the differences with limestone in the Maokou Formation.

8.4 Teaching Process and Precautions

8.4.1 Teaching process

(1) The teacher should remind students one day in advance to preview the Permian geological history and the stratigraphic sequence in South China.

(2) The teacher should briefly introduce the tasks, goals, and requirements at the starting point of the section.

8.4.2 Precautions

(1) The fossils in the bioclastic limestone (like the fusulinid) are small, thus every student and teacher need to bring a magnifying glass.

(2) Do not collect specimens in dangerous places.

(3) Bring water.

8.5 Focused Study and Reflections

(1) Analyze the assemblage characteristics and sedimentary microfacies of calcareous algae and foraminifera fossils in carbonate rocks.

(2) Analyze the influence of "the Dongwu Movement" on the environmental biota of the shallow-water platform in Middle and Late Permian.

9 ROUTE SEVEN: OBSERVING MIDDLE TRIASSIC–MIDDLE JURASSIC STRATIGRAPHY

9.1 Teaching Route

Zigui Base–Wenhua Village–Jinjigou Bridge–Wangjialing Tunnel–Zigui Base

9.2 Teaching Tasks and Requirements

(1) Understanding the stratigraphic sequences from the Middle Triassic Badong Formation to the Middle Jurassic Qianfoya Formation and gaining some preliminarily knowledge of the features of different clastic sedimentary rocks.

(2) Observing the features of parallel unconformity surface, basal conglomerate, and various sedimentary structures, and drawing field sketches.

(3) Measuring part of the Qianfoya Formation outcrops and sketching stratigraphic sections.

9.3 Route Information and Observing Points

The Triassic–Jurassic strata exposed in the Zigui area consists of the Middle Triassic Badong Formation, the Upper Triassic Jiuligang Formation, and the Lower Jurassic Tongzhuyuan Formation passing up to the Middle Jurassic Qianfoya Formation (Figure 9-1).

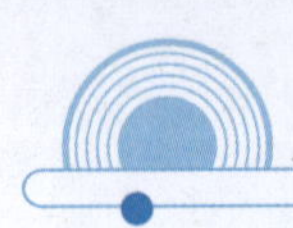

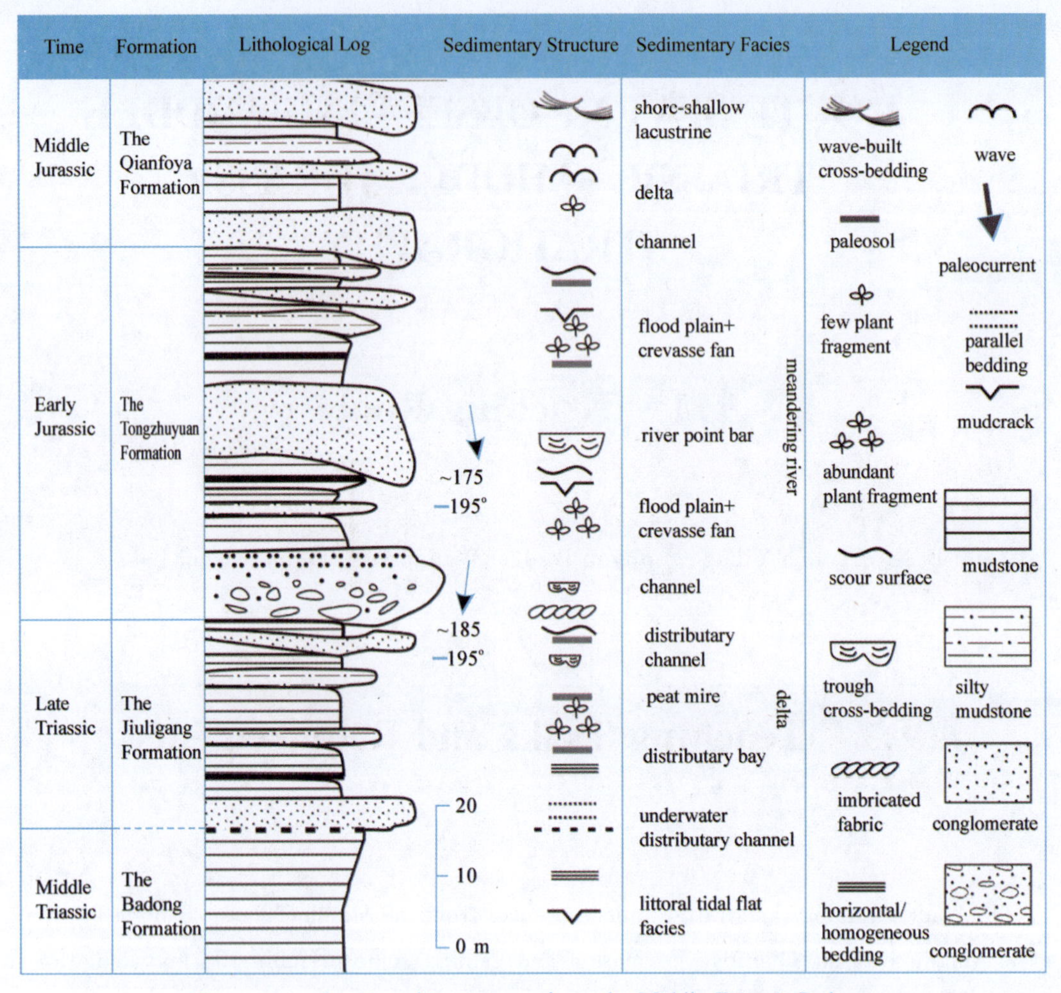

Figure 9-1 Stratigraphic sequence from the Middle Triassic Badong Formation to the Middle Jurassic Qianfoya Formation in Zigui, Hubei

9.3.1 The Badong Formation (T_2b)

The Badong Formation is mainly composed of redish-purple thick-bedded–massive mudstones, marlstones, and mottled medium–thin-bedded siltstones. Its bottom is distinguished from the Lower Triassic Daye Formation by brecciated limestones and marlstones. Observing the contact relationship between these two formations above is the main task for this route. Lithology observation and description are also required to be finished during the practice, despite the poor condition of the outcrop of the Badong Formation.

9.3.2 The Jiuligang Formation (T_3j)

The Badong Formation is overlain by the Jiuligang Formation which is dominated by a series of coal-bearing clastic rocks including greyish-black medium-bedded sandstones, siltstones, mudstones, and grayish-white oolitic aluminum mudstones with weathered siderite ooid, most of them have been weathered with rust mottles (Figure 9-2A). In addition, a large number of plant fossil fragments can be seen in shale.

The main fieldwork tasks include:

(1) Observing the contact relationship between the Jiuligang Formation and the Badong Formation.

(2) Observing the sedimentary sequences of the sandstones in the Jiuligang Formation.

Figure 9-2 The contact boundary between the Jiuligang Formation and the Badong Formation (A), the Tongzhuyuan Formation (B, C, and D)

9.3.3 The Tongzhuyuan Formation (J_1t)

The bottom of the Tongzhuyuan Formation is characterized by medium-bedded conglomerate, overlying the sandstones (Figure 9-2B) or black mudstone of the Jiuligang Formation. They contact in the erosional surface (Figure 9-2C). Shapes of gravels tend to be subangular–subrounded, compositions of gravels include chert, quartz sandstone, quartzite, vein quartz, etc. and the sizes of those range from 2 cm to 10 cm. The gravels are generally matrix-supported and partly grain-supported. Besides, the coarse-grained quartz sandstones fill the gap within gravels. The tubular gravels manifest imbricated fabric (Figure 9-2D), which can be used to construct the paleocurrent direction after the calibration of the corresponding strata attitude. The conglomerate layer is about 10-meter thick, successively passing upwards into lenticular interlayers within medium–thick-bedded sandstones.

Above the basal conglomerates, the Tongzhuyuan Formation is characterized by an alternation of medium–thick-bedded sandstones and black mudstones (Figure 9-3A), containing several coal seams (Figure 9-3B). In addition, abundant plant fossil fragments, referring to well-preserved fossils of ancient leaves and stems were discovered. The sedimentary structure of the sandstones is dominated by trough cross-beddings (Figure 9-3C and D).

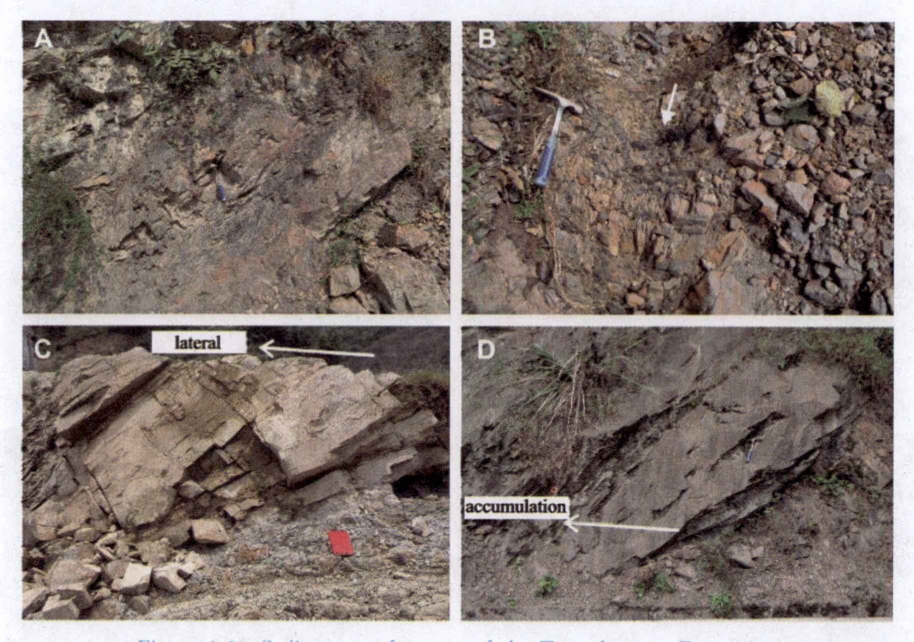

Figure 9-3 Sedimentary features of the Tongzhuyuan Formation

A. sandstone alternated with mudstone; B. coal seams in medium–thin-bedded mudstone-siltstone; C. erosional surface and lateral accumulation sequence at the bottom of channel deposits; D. lateral accumulation sequence

Structures of fluvial deposit and the sedimentary sequences are well-developed in the Tongzhuyuan Formation, including lateral accumulation of point bar, trough cross-beddings, channel scour surfaces, siltstone, and fine-grained sand deposits of crevasse fan, mudstone of flood plain, etc. The main fieldwork tasks include:

(1) Observing the erosional contact in the boundary between the Tongzhuyuan Formation and the Jiuligang Formation.

(2) Observing the composition, structural features, and paleocurrent indicators of conglomerate at the bottom of the Tongzhuyuan Formation.

(3) Observing the fluvial sedimentary structures and sedimentary sequences of the Tongzhuyuan Formation.

9.3.4 The Qianfoya Formation (J_2q)

The Qianfoya Formation conformably overlays the Tongzhuyuan Formation. The bottom of the former appears to be the interbedding of slim gray sandstones and gray mudstones. The erosional surface is developed in the medium-bedded sandstones. Apart from those, black mudstones and medium–thin-bedded sandstones featured by symmetrical ripple marks (Figure 9-4A). The upper of this formation is mainly comprised of purple mudstones, siltstones, and grey sandstones, within very thick–thick-bedded sandstone, cross-beddings are well developed (Figure 9-4B). The Qianfoya Formation indicates lacustrine facies, including shore and shallow lake sandstones, and purple mudstones.

The main fieldwork tasks include:

(1) Observing the sandstone-mudstones of shore-shallow lacustrine facies.

(2) Observing the wave-built cross-bedding.

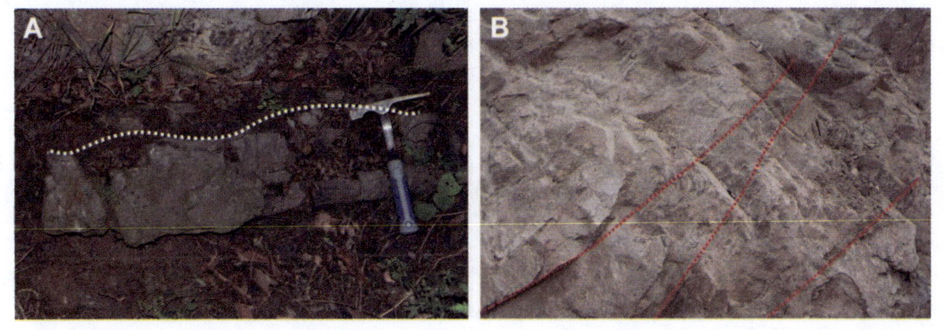

Figure 9-4 The sedimentary features of the Qianfoya Formation

A. symmetrical ripple marks; B. cross-beddings

9.4 Teaching Process and Precautions

9.4.1 Teaching process

(1) Teacher reminds students to preview the Triassic–Jurassic geological history and the stratigraphic sequence of South China.

(2) Teacher briefs the tasks, objectives, and requirements of the fieldwork at the starting point of the section. (5 minutes)

(3) Teacher introduces the Middle Triassic–Jurassic stratigraphic sequences and geological evolution of South China, especially the northern margin of the Yangtze Block. (10–20 minutes)

(4) Teacher introduces one observation point of the Middle Triassic Badong Formation.

(30 minutes)

(5) Teacher introduces two observation points of the Lower Jurassic Tongzhuyuan Formation. (60 minutes)

(6) Teacher introduces two observation points of the Middle Jurassic Qianfoya Formation. (60 minutes)

9.4.2 Precautions

(1) Bring measuring line, tape, and compass because it is required to measure the section on site.

(2) Do not collect samples in hazardous spots.

(3) Bring food and drinking water.

9.5 Focused Study and Reflections

(1) Analyze the detrital sequence of delta-fluvial-lacustrine deposits and sedimentary structural features.

(2) Discuss how the Triassic–Jurassic sedimentary records along the northern margin of South China link with the orogenic movement and paleoclimate change.

(3) Explore the age and pattern of the transition from carbonate to clastic sediments of South China and its linkage of the Indosinian Movement.

(4) Since the lithofacies assemblages of the Triassic–Jurassic sedimentary sequence vary widely, explain how the corresponding sedimentary environment evolved.

10 ROUTE EIGHT: OBSERVING STRUCTURAL GEOLOGY AND CAMBRIAN-ORDOVICIAN STRATA IN QINGJIANG, CHANGYANG

10.1 Teaching Route

Zigui Base-South Baishiqiao Bridge, Changyang-Xiaojiatai-South Baishiping Village-Zigui Base

10.2 Teaching Tasks and Requirements

(1) Observing and describing the Cambrian-Ordovician stratigraphic sequences.

(2) Learning the basic methods to observe fault and fold.

(3) Observing and describing the structural development along the south of Baishiqiao Bridge-Xiaojiatai Village-Baishiping Village, including: understanding Cambrian-Ordovician fold type and fracture structure development in the northern limb of Changyang anticline, basing on the systematic changes of the emergence stratum and stratigraphic attitude; observing and analyzing the influence of rock competence and layer thickness on fold form; analyzing the sequence of structural development; and sketching a profile of the route.

10.3 Route Information and Observing Points

The route is located at about 2 km east of Changyang, from south to north, starting from the north bank of the Qingjiang River and passing through Xiaojiatai to Baishiping Village. It is located in the northern limb near the east–west anticline in Changyang. The direction of structure line is near EW, and the primary stress is near SN.

We can observe the Upper Sinian Dengying Formation(Z_2dy), the Middle Cambrian Tianhcban Formation ($€_2t$), the Lower Cambrian Qinjiamiao Formation ($€_3q$), the Upper Cambrian–the Lower Ordovician Loushanguan Formation ($€_3O_1l$), the Lower Ordovician Nanjinguan Formation (O_1n), the Lower Ordovician Fenxiang Formation (O_1f), the Lower Ordovician Honghuayuan Formation (O_1h), and the Middle–Lower Ordovician Dawan Formation ($O_{1-2}d$).

The main structural features along the route are mainly folds and faults (Figure 10-1). The folds are box-shaped isochore folds that strike generally in an east-west direction, and the anticlines and synclines are arranged in a continuously parallel line. The syncline is gently wide-opened and the anticline is relatively tightly closed, the overall form is similar to that of Jura-type folds. There are various types of faults, including longitudinal faults and oblique faults in relation to the direction of regional structure lines. Strike-slip faults, reverse faults, and normal faults are developed according to the slip motion. Fold structures may be mainly developed in Late Triassic–Early Jurassic, while fault structures are the products of multi-stage tectonic movements during Indosinian, Yanshanian, and Himalayan.

10.3.1 The Baishiqiao fault on the south side of Baishiqiao Bridge and Liangpan Series of small structural observation points

10.3.1.1 *Observation of the Baishiqiao fault*

The Baishiqiao fault appears as a 1–4 m wide fracture zone (Figure 10-2). The overall attitude of the fracture zone (F) is 230°∠70°, 245°∠65°; the slickenside attitude on the main fault surface (L_a) is 150°/10°. There are large structural lens composed of flattened dolomites in the zone; folded zones composed of thin-bedded muid-ribbon limestone; partially visible variegated breccia-cataclasite rock zone composed of yellow, gray, and brown structural breccia, granular rock, and broken siltstone(Figure 10-3). The hanging wall is the light gray thick-bedded–massive dolomite of the Sinian Dengying Formation (Z_2dy), the stratum attitude (S_0) is 354°∠67°, with occasional thin-bedded dolomite limestone. The footwall is the Cambrian Tianheban Formation ($€_2t$) light gray and dark gray thin-bedded muid-ribbon limestone intercalated with medium bedded limestone, and the attitude of rock stratum (S_0) is 355°∠67°. The missing strata between the two include the Cambrian Yanjiahe Formation ($€_1y$), the Shuijingtuo Formation ($€_2s$), and the Shipai Formation ($€_2sh$). The Baishiqiao fault is a sinistral strike-slip fault (Figures 10-4 and 10-5) judging from the attitudes of main faults and slickensides, the steps on the main fault surface, ect.

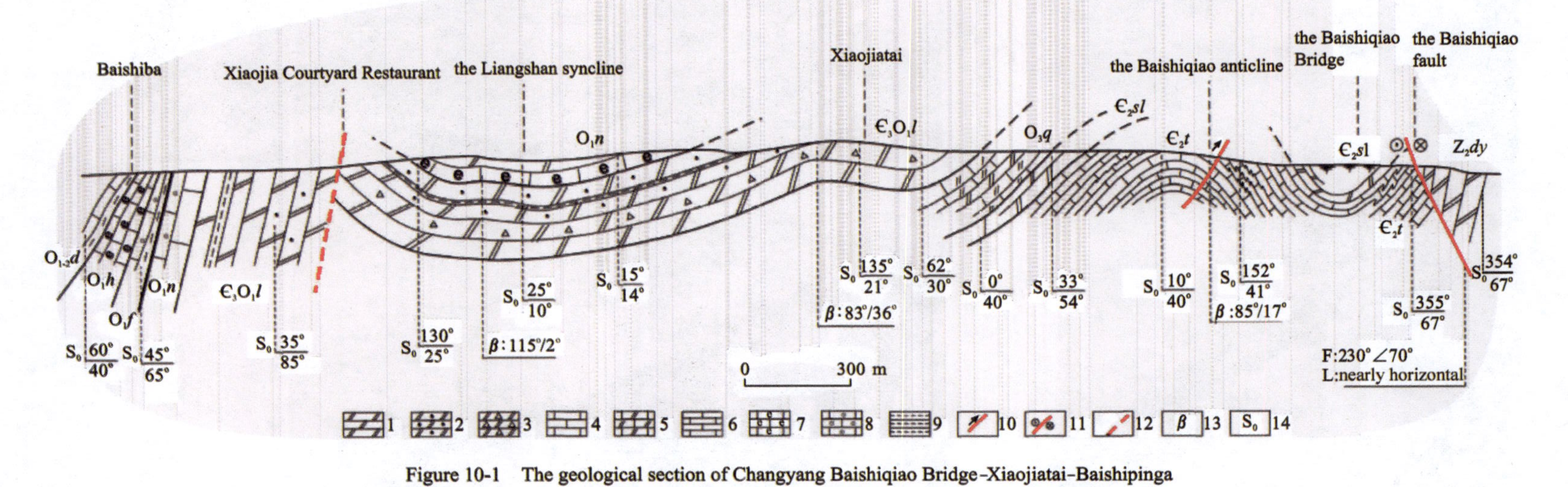

Figure 10-1 The geological section of Changyang Baishiqiao Bridge-Xiaojiatai-Baishipinga

1. dolomite; 2. sandy dolomite; 3. brecciated dolomite; 4. limestone; 5. dolomitic limestone; 6. thin-bedded banded limestone; 7. bioclastic limestone; 8. oolitic limestone; 9. thin-bedded mudstone; 10. reverse fault; 11. wrench fault; 12. infer fault character unknown; 13. fold hinge; 14. bedding.

Figure 10-2　Sinistral strike-slip fault at Baishiqiao Bridge, Changyang, Hubei

Figure 10-3　The variegated structure/tectonic breccia in the fracture zone of the Baishiqiao fault

(pay attention to the nearly horizontal slickensides on the main fault plane)

Figure 10-4　Bedding brittle-ductile shear zone in the Dengying Formation

(the deformation zone cuts the early calcite-filled veins)

Figure 10-5 The approximately horizontal slickenside and step on the main fault of the Baishiqiao fault, indicating sinistral strike-slip movement

10.3.1.2 *Observation of bedding shear deformation structure in the Dengying Formation of the southwest wall of the Baishiqiao fault*

The southwest wall of the Baishiqiao Fault is light gray thick-bedded–massive dolomite of the Sinian Dengying Formation (Z_2dy), and the rock attitude (S_0) is 354°∠67°. A series of bedding shear foliations are developed, and a bedding brittle-ductile shear zone with a width of about 30 cm can be seen (Figure 10-6). The lineation on the shear foliation is nearly horizontal and the slickensides can be measured, the attitude of slickensides (L) is 298°∠6°. According to the near-horizontal slickensides and normal steps, it is reflected as a left-lateral shear sliding in bedding (Figures 10-6 and 10-7).

Figure 10-6 The approximately horizontal slickensides and normal steps on the bedding brittle-ductile shear zone of the Dengying Formation, indicating sinistral slip

Figure 10-7 The approximately horizontal slickensides on the wavy large bedding shear slip surface in the Dengying Formation

The bedding shear foliations are cut by the Baishiqiao fault, indicating they formed earlier than the development of the Baishiqiao fault.

10.3.1.3 *Observation of joint structure in the Dengying Formation at the southwest wall of the Baishiqiao fault*

The joint structure in the light gray thick-bedded–massive dolomite of the Dengying Formation (Z_2dy) is well developed and filled with veins of calcite. The forms of calcite veins are parallel, echelon-shaped, and torch-shaped, reflecting the products of a unified tectonic stress field (the attitude of various veins can be measured, the mechanical properties of joints can be interpreted, and could be analyzed the tectonic stress field) (Figures 10-8 and 10-9).

Figure 10-8 Calcite-vein-filled joint structures in the dolomites of the Dengying Formation, arranged in parallel or echelon

Figure 10-9 Torch-shaped joints filled with calcite in the dolomites of the Dengying Formation

The series of veins are cut by the bedding shear zone, reflecting that they formed earlier than the bedding shear zone. However, these veins are also cut by a set of nearly north–south high angle shear joints. Therefore, their formation should be earlier than the nearly south–north shear joints (Figure 10-10).

Figure 10-10 Fracture system filled with early calcite cut by subsequent nearly NS-trending shear joints

10.3.1.4 *The asymmetric fold structure in the Tianheban Formation in the northeast wall of the Baishiqiao fault*

The Tianheban Formation ($\epsilon_2 t$) in the northeast wall of the Baishiqiao fault is a set of light gray and dark gray thin bedded mud-ribbon limestone interbedded with medium bedded limestone; the attitude of the main strata (S_0) is 355°∠67°. Under the influence of bedding shear deformation, a series of asymmetric fold structures are developed. The deformation indicated by the asymmetric fold is reverse bedding shear slipping. Under the influence of bedding shear slipping, some competent rock layers are deformed into lens.

10.3.2 Observation point of the Baishiqiao anticline on the north side of the Baishiqiao Bridge

This observation point reveals an upright plunging fold (Figure 10-11) composed of light and dark gray thin-bedded argillaceous limestone in the Cambrian Tianheban Formation ($\epsilon_2 t$) sandwiched with medium–thick-bedded limestone. The fold hinge zone of the cylindrical competent layer is composed of medium–thick-bedded limestone. The altitude of the south wing (S_0) is 152°∠41°, the altitude of the north wing (S_0) is 10°∠40°. Altitude of hinge line (L_b) is 85°∠19°. The axial plane is nearly east–west, nearly upright. The thin-bedded mud-ribbon limestone above and below the main layer is a incompetent layer with a series of parasitic folds. A large number of S-shaped parasitic folds and compound S-shaped parasitic folds (Figure 10-12) are seen in the south wing; Z-type parasitic folds (Figure 10-13) and kink bands are developed in the north wing, box folds formed by bedding slipping and conjugate kinks associated with them can also be seen (Figure 10-14).

Figure 10-11 The Baishiqiao anticline

Figure 10-12 Typical S-shaped fold (left) and compound S-shaped fold (right) in the south limb of the Baishiqiao anticline

Figure 10-13 Z-shaped parasitic folds in the north wing of the Baishiqiao anticline

Figure 10-14 Box folds and conjugate kinks formed by the bedding slipping of thin-bedded limestone in the Cambrian Tianheban Formation in the north limb of the Baishiqiao anticline

10.3.3 Observation of the Mid–Late Cambrian stratigraphic unit sequence and detachment-type bedding faults at the observation point on the north side of the Baishiqiao anticline

10.3.3.1 *Conformity contact observation point of the Tianheban Formation and the Shilongdong Formation*

The Tianheban Formation ($\in_2 t$) gray thin-bedded limestone interlayered with medium thick-bedded limestone assemblage is exposed in the south of the point. The fresh surface of the rock is dark gray, with fine-grained structure and thin–medium thick-bedded structure. The single layer thickness is 30–50 cm. The major mineral component is microcrystalline calcite, accounting for about 95%, with a little argillaceous and silty. The rock has conchoidal fracture and is compact, which can be carved by a knife. Z-type parasitic folds are developed in thin-bedded mud-ribbon limestone.

The Shilongdong Formation ($\in_2 sl$) is exposed on the north side of the point. The formation is composed of dark gray–brown medium–thick-bedded dolomite and dark gray thin-bedded dolomite. The fresh surface of rock is dark gray–brown with aplitic texture and thin–medium thick-bedded structure. The major composition is microcrystalline dolomite, which accounts for 90%.

The two sets of strata are in conformable contact and the attitude of the bedding surface is 0°∠40°. The thickness of stratum varies from thick to thin to thick in the north direction along the route.

Along the mian road 30 m northward, it is a combination of dark gray–brown medium–thick-bedded dolomite and dark gray thin-bedded dolomite of the Shilongdong Formation ($\epsilon_2 sl$), with an attitude (S_0) of 15°∠50°.

10.3.3.2 *Conformable contact observation point of the Shilongdong Formation and the Qinjiamiao Formation*

The Shilongdong Formation ($\epsilon_2 sl$) on the south side is composed of dark gray–brown medium–thick-bedded dolomite and dark gray thin-bedded dolomite.

The Qinjiamiao Formation ($\epsilon_3 q$) on the north side is gray–dark gray thin-bedded dolomite, as well as dolomitic limestone with a small amount of mudstones.

The Shilongdong Formation and the Qinjiamiao Formation are in conformable contact.

Continue to walk along the main road to the north for 100 m, there are gray–dark gray thin-bedded dolomite and dolomitic limestone of the Qinjiamiao Formation ($\epsilon_3 q$), interlayered with a small amount of mudstones. The attitude of bedding (S_0) is 4°∠49°.

A series of bedding detachment faults are developed in the gray and dark gray thin-bedded dolomite and dolomitic limestone of the Qinjiamiao Formation ($\epsilon_3 q$). The attitude of fault zone is nearly parallel to the bedding. The width of a single slip band varies from a few centimeters to more than ten centimeters. The thinning and missing of strata is caused by slip deformation (Figure 10-15).

10.3.3.3 *The boundary between the Qinjiamiao Formation and the Loushanguan Formation (300 m away from the Baishiqiao anticline observation point to the north)*

On the south side of the Qinjiamiao Formation ($\epsilon_3 q$) is gray–dark gray thin-bedded dolomite and dolomitic limestone with gray weathered surface and dark gray fresh surface; fine-crystalline structure, medium–thin-bedded structure. It is mainly composed of microcrystalline calcite, with a content of about 80%, and a dolomite content of about 15%, with chop profile occasionally seen locally. The attitude of bedding (S_0) is 4°∠49°.

The Loushanguan Formation ($\epsilon_3 O_1 l$) on the north side is a combination of light gray thick-bedded dolomite and breccia dolomite. The fresh surface of light gray thick-bedded dolomite is light gray with aplitic texture and thick-bedded structure. The major composition is microcrystalline dolomite, which accounts for 90%. Breccia dolomites are mostly angular and sub-angular, with calcium cementation (Figure 10-16).

The Qinjiamiao Formation ($\epsilon_3 q$) and the Loushanguan Formation ($\epsilon_3 O_1 l$) are conformable in contact.

10.3.4 Observation of Liangshan composite syncline structure

An open syncline structure is developed between Xiaojiatai and Xiaojia Courtyard

Figure 10-15 Strata thinning and missing due to the northward bedding forward detachment

Figure 10-16 Breccia dolomite of the Loushanguan Formation (Ge Mengchun et al., 2003)

Restaurant. The core stratum is the thick-bedded limestone and limestone dolomite of the Lower Ordovician Nanjinguan Formation (O_1n). The two wings are symmetrically exposed with thick-bedded dolomites from the Upper Cambrian to the Lower Cambrian; the dolomite is the Ordovician Loushanguan Formation (ϵ_3O_1l) thick-bedded dolomite. The folds are generally open and gentle. The representative stratum attitude in the south wing (S_0) is 25°∠10°, the representative stratum attitude in the north wing (S_0) is 130°∠25°, the representative stratum attitude at the hinge zone is 115°∠2°, indicating an overall vertically horizontal fold.

The interior of the composite syncline is complicated by a series of minor wavy folds, e.g., in the south limb at Xiaojiatai, minor folds are developed. On the south side is the Xiaojiatai minor syncline consisted of the Loushanguan Formation. The syncline core is composed of the upper Loushanguan Formation and the two limbs are composed of the lower Loushanguan Formation. The attitude of the south limb (S_0) is 50°∠25°; the attitude of the north limb (S_0) is 135°∠21°; the attitude at the hinge zone is 90°∠17°, representing the attitude of the hinge line. The axial plane is nearly upright in the east-west direction, which is a gentle and open upright plunging syncline.

On the north limb, the Xiaojiatai minor anticline is developed, and the lower Loushanguan Formation constitute the core of the anticline. The upper Loushanguan Formation constitute the south limb of the anticline. The attitude of the south limb (S_0) is 135°∠21°; the upper Loushanguan Formation and the Nanjinguan Formation form the north wing, and the stratum attitude of the north wing (S_0) is 30°∠42°. The attitude at the hinge zone is 85°∠36°, which represents the high line attitude. The axial plane is upright, nearly east-west, which is a gentle and open upright plunging anticline.

10.3.5 Observation and description of the Late Cambrian-Early Ordovician stratigraphic sequences in the southern part of Baishiping Village

In about 600 m, steeply dipping stratigraphic sequences occur. The overall strata are dipping to the north at a high angle. Several Early Ordovician lithostratigraphic units are developed here. The strata tend to be younger from the south to the north, and sequentially outcrop the Upper Cambrian-the Lower Ordovician Loushanguan Formation (ϵ_3O_1l), the Lower Ordovician Nanjinguan Formation (O_1n), the Fenxiang Formation (O_1f), the Honghuayuan Formation (O_1h), and the Middle-Lower Dawan Formation ($O_{1-2}d$).

The Upper Cambrian-the Lower Ordovician Loushanguan Formation (ϵ_3O_1l): gray thick-layered dolomite and sandy dolomite.

The Lower Ordovician Nanjinguan Formation (O_1n): gray thick-layered limestone, breccia limestone, calcite dolomite, and oolitic limestone.

The Lower Ordovician Fenxiang Formation (O_1f): gray medium-thin-bedded bioclastic limestone, oolitic limestone sandwiched with mudstone and shale with fossils of tongue-shaped shell.

The Lower Ordovician Honghuayuan Formation (O_1h): gray medium-thin-bedded bioclastic limestone.

The Middle–Lower Ordovician Dawan Formation ($O_{1-2}d$): gray–greyish green medium-thick-bedded argillaceous limestone interlayered with greyish-green mudstone and shale. Nodular limestone containing Yangtzeela fossils is developed in the area.

Wide Quaternary coverage occurs between the steeply dipping stratigraphic belt and the gentle and open Liangshan anticline in the south, but in macroscope, these show structural inconsistency with a sudden change in occurrence. A sudden change in bedding from gentle on the south side to steep and vertical in the southeast. It is speculated that there is a large-scale fault in between, and the landscape appears as a linear valley near east–west direction (Figure 10-1).

10.4 Teaching Process and Precautions

10.4.1 Teaching process

(1) Teachers should remind students to preview the Sinian–Ordovician stratigraphic sequences and to review the concepts of Jura-type folds, parasitic folds, and the collection of geometric elements of fold and fault one day in advance.

(2) At the starting point of the profile, the teachers briefly introduce the tasks, goals, requirements, the characteristics of regional decollement structures, and the folds of the route profile in the north limb of the Changyang anticline.

(3) Before noon, complete the sketch of the Baishiqiao fault and the Baishiqiao anticline, the data collection of attitude, and the observation and description of the Cambrian stratigraphy. After lunch, the observation and description of the syncline structure and the Ordovician stratigraphy should be completed.

10.4.2 Precautions

(1) Students should be reminded to pay attention to safety as the route is along the main traffic road.

(2) Students are required to diaw a section sketch of the route and measure continuously the attitudes of strata.

(3) Focuse on training how to analyze structures, especially the methods of analyzing the structural association and its stage matching.

(4) Lunch and drinking water are required as it will take about 1.5 hours to reach the site.

10.5 Focused Study and Reflections

(1) Analyze the relationship between the lithology, stratum thickness, and the rock competence, as well as the influence of rock competence on strength and deformation.

(2) Observe fold structure style and quantitative statistical analysis (β and π graphic methods).

(3) Explore fault structure association and fault stage division.

(4) Discuss structural association and sequence division of deformation.

11 ROUTE NINE: OBSERVING THE XIANNVSHAN FAULT AND RELATED STRUCTURES

11.1 Teaching Route

Zigui Base–Cuijiaping–Huangkouping–Zhouping–Zigui Base

11.2 Teaching Tasks and Requirements

(1) Understanding the lithological characteristics and sedimentary environment background of representative strata in Mesozoic.

(2) Mastering the fieldtrip methods of observation and analyzing fault structures and related structures of geometry and kinematics.

(3) Making a profile map of the fault structures.

(4) Analyzing the mechanical background formed by fault structures.

11.3 Route Information and Observing Points

The Xiannvshan fault (Fairy Mountain Fault) is a large-scale and representative fault structure in the Three Gorges area of the western Hubei Province. The fault structure is located in the southwest of the Huangling Dome structure (quaquaversal structure), starting from Yuyangguan–Qingshuiwan on the south side, and ending in the Xiling Gorge area of the Yangtze River. It extends in the direction of NNW and extends approximately 90 km in strike. The overall strike is stable, and the dip angle is generally high or nearly vertical. This route covers the northern section of the Xiannvshan fault, with good traffic conditions along the route, and has crossed the Xiannvshan fault structure many times.

The strata exposed along this route mainly include the Lower Silurian Xintan Formation (S_1x), the Middle–Lower Triassic Jialingjiang Formation ($T_{1-2}j$), the Lower Triassic Daye Formation (T_1d), and the Lower Cretaceous Shimen Formation (K_1s).

The route has typical geological structures, and the main fault structures and various associated fold structures are well developed. The fold structure is dominated by east–west folds, but also has a north–south direction, and the two are structurally superimposed. The fault structure is dominated by the NNW-trending Xiannvshan fault system. The fold and fault structures mainly formed during Yanshanian.

11.3.1 Asymmetric fold and mullion structure in Cuijiaping

This point is located at the northern end of the Xiannvshan fault. In terms of geomorphology, it is a deep-cut and trough-like landform, but the deformation effect of the fault is relatively weak, showing only a set of near-upright fault plane in the NNW–SSE direction (Figure 11-1).

The lithology of the Jialingjiang Formation is gray medium–thick-bedded limestone. The attitude (S_0) is stable: 263°∠34°. At this point, asymmetrical folds near the north–south direction can be seen in the quarry, and the fold hinges are nearly horizontal and Z-shaped, indicating a slippage effect to the west (Figure 11-2). The attitude of hinge (L) is 341°∠14°. At the same time, the EW fold of the hinge can be seen at this point (Figure 11-2), and the attitude of the fold hinge (L) is 251°∠18°, indicating the extrusion stress of north–south direction.

Figure 11-1 High-angle fold of the Jialingjiang Formation in the Cuijiaping quarry (lens to the south)

Figure 11-2 Asymmetric folds in the north–south direction of the Cuijiaping quarry (positive sliding, lens to north)

The cliff strata on the northwest road side of the quarry is gray thick-bedded limestone, partially laminar limestone, with a series of mullion structures (Figure 11-4), and the anticline presents a wide and gentle arc-shaped bulge with sharp edges in between. The syncline of the two is stable and arranged regularly. The hinge inclines to the west with an attitude (L) of 268°∠39°.

The elements of the field observation of the mullion structure include the pivot attitude, the wavelength of the mullion structure, the wave amplitude, and the thickness of the rock layer. The above-mentioned data are the basic data for strain measurement and analysis of rock competence. At the same time, pay attention to guiding students to observe various associated structures in different parts of the mullion structure, including the characteristics of plastic deformation and brittle deformation, and to compare and analyze the competence and origin of rocks.

Figure 11-3 Folds of east–west direction at the Cuijiaping quarry (lens to east)

Figure 11-4 Mullion structure at Cuijiaping quarry (lens to the southwest)

11.3.2 Observation point of the oblique thrust tectonics of Huangkouping (South) Xiannvshan fault

Here is the observation point of the Xiannvshan fault. The fault plane is inclined to the west at a high angle. The footwall (east panel) of the fault is the Lower Cretaceous Shimen Formation (K_1s), which is purplish-red massive cobblestone with glomerate sandstone. The gravel is well rounded, medium–poor sorted, muddy, and sandy. The attitude (S_0) is 278°∠35°.

The hanging wall (west panel) of the fault is the Lower Triassic Daye Formation (T_1d), which is gray–grayish-black and thick-bedded limestone. The two are in contact with faults (Figure 11-5), and the fault zone is about 50 cm. Lenses, fault cleavage, and asymmetric folds are developed in the fault zone. Clear oblique fault scratches and steps are seen on the limestone section of the Daye Formation (Figure 11-6), indicating a rightward oblique

thrust. The attitude of fault plane (F) is 260°∠75°, scratch lineation (L) is 173°∠39°.

Figure 11-5 The Xiannvshan fault at Huangkouping (lens to the southwest)

Figure 11-6 The scratch step structure of the Xiannvshan fault at Huangkouping (a rightward oblique thrust, lens to the west)

11.3.3 Strike-slip structure observation point of the Xiannvshan fault at Zhouping

This point is located in the northeast of Zhouping, and the Xiannvshan Fault crosses this point, forming obvious negative topography (Figure 11-7). On the west side, there is the Lower Silurian Xintan Formation (S_1x), which is green medium–thin-bedded sandstone/argillaceous sandstone (Figure 11-8), with a steep attitude (S_0) of 20°∠83°.

To the east of the point, there is the Lower Cretaceous Shimen Formation (K_1s), which is light red–carmine and grayish-brown thick-bedded gravel-bearing sandstone and sandstone. The gravels are well rounded, and the sorting is average–poor. The attitude of cardium sandstone is 189°∠12°. The oblique bedding is developed in the sandstone, and the oblique bedding occurrence (Scb) is 72°∠40°.

Figure 11-7 The faulted landform of Xiannvshan at Zhouping (lens to the south)

Figure 11-8 The Silurian Xintan Formation (lens to the east)

The two formations are in contact with faults, and the fault band is several meters wide. They are broken zones composed of glutenite and large limestone lens of the Shimen

Formation. The glutenite fault zone of the Shimen Formation has developed fault cleavage (Figure 11-9), with developed lenses and gentle scratches on the fractured surface of the fault (Figure 11-10). Calcite developed along the fault plane and developed in steps, indicating rightward/dextral shear.

The fault attitude (F) is 80°∠76°, scratch attitude is 178°∠24°. The Cretaceous on the east side develops a soothing anticline.

Figure 11-9 The fault zone of the Xiannvshan fault at Zhouping (lens to the north)

Figure 11-10 The scratch step structure of the Xiannvshan fault zone at Zhouping (dextral shear, lens to the west)

11.4 Teaching Process and Precautions

11.4.1 Teaching process

(1) Students preview the observation methods of Mesozoic strata and faulted structures in the study area, understand the method of measuring with a compass and the method of observing lines and structures, and make full preparations.

(2) Teacher introduces the tasks, goals, and requirements of the fieldtrip route at the Starting point. (5 minutes)

(3) Teacher introduces the observation points at Cuijiaping. (60 minutes)

(4) Teacher introduces the observation points at Huangkouping. (45 minutes)

(5) Teacher introduces the observation points at Zhouping. (45 minutes)

11.4.2 Precautions

(1) Pay attention to the safety of fieldtrip teaching because the geological site is along the

provincial highway, and the safety observers are needed.

(2) Preview and be proficient in the method of measuring with a compass and the method ofobserving lines and structures. Bring a ruler (or a steel tape).

(3) Bring water and lunch.

11.5 Focused Study and Reflections

(1) Competence comparison and strain measurement analysis of mullion structure.

(2) Fracture properties, mechanical state, and formation time of the Xiannvshan fault.

(3) Modern activity and crustal stability of the Xiannvshan fault.

(4) Is the cutting of the Xiannvshan fault deeply involved in the basement?

12 ROUTE TEN: THE QUATERNARY ROUTE

12.1 Teaching Route

Zigui Base–Huangniuyan–Heshangdong–Guancaiyan–Datang–Zhangjiachong–Zigui Base

12.2 Teaching Tasks and Requirements

(1) Learning and understanding the Three Gorges Project.

(2) Learning and understanding the regional geomorphological features and the Quaternary sediments.

(3) Observing the gravity geological disasters such as collapse and faulted landslides, analyzing their causes, learning the basic engineering approaches in the prevention and control of potential unstable rock body.

(4) Understanding the scouring effect of surface flow on slope soil and observing the basic content and approaches in water and soil conservation in watershed.

12.3 Route Information and Observing Points

On the Quaternary route, we will observe the landscape in the Three Gorges Dam area and the Yangtze River valley, and learn to investigate and describe the gravitational geomorphology, including the karst, the cave deposit, the collapse, and the faulted landslides, as well as deposits, such as colluvium, diluvium, and eluvium. On the route, we can observe the landscape of the internship area from a distance, and study the Quaternary deposits formed for various causes in a closer view, which belongs to the Quaternary route with diversified geomorphological characters.

12.3.1 The Yangtze River Valley and the planation surface in Huangniuyan

12.3.1.1 *Observation and reflection of the man-made landscape in the Three Gorges Dam area*

1. Brief introduction of the Three Gorges Dam Project

The Three Gorges Dam Project is the world's largest hydroelectric power station. The dam has a height of 185 m and a width of 2,335 m. The top of the dam is about 40 m in width, and the bottom is 115 m in width. The Three Gorges Dam covers a reservoir area of about 1,084 km^2 and a catchment area of about one million square kilometers. The normal water level of the Three Gorges Dam is about 175 m, and its total storage volume is about 39.3 billion cubic meters. The project was officially launched in December 1994 and was built up in July 2012, with power generation started in July 2003.

2. Performance of the Three Gorges Dam Project

(1) Flood prevention and control. After the Three Gorges Dam was built up, the reservoir has remained a normal water level at about 175 m, with reserved flood control capacity at about 22.15 billion cubic meters. The Three Gorges hydrojunction has a maximum flood discharge capacity of about 116,110 m^3/s and can cut the peak discharge by 27,000–33,000 m^3/s, which makes it the largest water project worldwide. Its massive regulation and storage capacity makes it possible for the Jianghan Plain to avoid mass casualties and catastrophic disasters when stricken by catastrophic flood.

(2) Power generation. A total of 32 hydroelectric generating sets are installed, with a unit capacity of 700,000 kW (including the right bank underground power station, which contains six hydroelectric generating sets with a unit capacity of 700,000 kW) and a gross installed capacity of about 22.5 million kW. Its annual maximum generating capacity stands at about 100 billion kW • h, making it the world's largest hydropower station.

(3) Shipping. The Three Gorges of the Yangtze River is characterized by narrow surfaces, rushing currents, and wide-spread submerged reefs, which makes it even harder for boats to sail upwards. After the Three Gorges Dam had been built up, the Yangtze River has been widened to become lakes in this part, which reduces the flow and enables travelling by ships with ten-thousand-tons capacity from Shanghai directly to Chongqing. Moreover, the

shipping conditions of the lower-middle reaches of the Yangtze River are also improved by discharging water from the reservoir during low flow season.

(4) Another important function of the Three Gorges Dam Project is to combat drought and serve as ecological water supplement in the lower–middle reaches of the Yangtze River.

12.3.1.2 *The Three Gorges Dam's impacts on the geological process of the Yangtze River and its resource environment effect*

The Three Gorges Dam by cutting down the Yangtze River, is bound of induce huge impact on resources and environments of the Yangtze River by changing the geological processes of the Yangtze River. In this case, teachers and students can discuss about the impacts. If time is limited in the field, students can be assigned to think about the question on their own. The outline is as follows:

(1) The impacts of the Three Gorges Dam reservoir on the surrounding crustal stress field and its resource and environmental effects.

(2) The impacts of the Three Gorges Dam reservoir on the slope geology of the Three Gorges area and its resource and environmental effects.

(3) The Three Gorges Dam reservoir's impacts on the microclimate effects of the region and its resource and environmental effects.

(4) The impacts of the Three Gorges Dam reservoir on the geological process of its upper reach of the Yangtze River and its resource and environmental effects.

(5) The impacts of the Three Gorges Dam reservoir on the geological process of the middle–lower reaches of the Yangtze River and its resource and environment effects.

(6) The impacts of the Three Gorges Dam reservoir on the geological process in the estuary of the Yangtze River and its resource and environmental effects.

12.3.1.3 *Observation of the landscape of the region*

Two macro-landscapes in the region can be observed in this site, namely the planation surface and the wide valley (or erosional terrace).

1. Planation surface

Looking westward from the site, we can observe an array of peaks and ridges with almost the same height at about 1,000 m above the sea level which represent the planation surface of the Three Gorges stage in the Three Gorges area.

Davis divided the geomorphologic evolution into three stages, notably, the infancy stage, the maturity stage, and the old stage. The peneplain formed during the old stage can be re-eroded either due to the crustal uplift or decrease of the base level of erosion. The remaining series of flat hilltops are almost at the same height. Such geomorphic surface depicts the appearance of the peneplain during geological stage, which is called the planation surface. Multi-phase uplifts lead to multi-phase planation surfaces, which are numbered from top to bottom one by one, opposite to that of fluvial terrace.

Since the Himalayan Movement during Cenozoic, the Three Gorges area has been subjected to intermittently tectonic uplifts on the large-scale, forming a third-phase planation surfaces in the area (Xie et al., 2006). The remains of the first session, i.e., the Exi session, are scattered in the watershed area of the Yangtze River and the Qingjiang River, which is about 1700–2000 m above the horizon. The second session, or the Shanyuan session, is widely seen on the beaches of the Yangtze River and its branches. The third session, or the Three Gorges stage, is about 1000 m above the ground. Viewing from the observation platform, we can see that the mountains in the farthest west and those of the farthest east are almost at the same sea level of about 900–1000 m, which is the third session planation surface in this area.

2. Wide valley

Looking downward from the left, we can see four platforms with approximately the same height at about 650 m, which is the wide valley (Figure 12-1). The wide valley is quite common on both sides of the Three Gorges area of the Yangtze River.

Figure 12-1 The landscape of the planation surface and wide valley

12.3.2 Karst landscape and cave deposit at Heshangdong

12.3.2.1 *Architecture of the cave and its causes*

The Heshangdong is situated on the brink of the karst depression, which extends 250 m in length and 100 m in width. During high flow period, the surface water intermittently flows into a sinkhole alongside the eastern edge of the depression and trickles down via infiltration whilst surficial runoff is absent at ordinary times.

The entrance of the cave presents as a triangle, which is 40 m in height and 20 m in width. Going inward, we can see that the space expands slightly, with a depth of about 50 m. The cave is formed within the Sinian medium–thick-bedded dolomite of the Hamajing Member of the Dengying Formation (Z_2dy^h). The rock here is soluble, but its solubility is second to typical limestone cave. Therefore, apart from the stalactite on the roof of the cave, limited stalagmite on the ground indicates the unsaturated infiltrated karst water. The bottom of the cave and the depressions are marked by the fourth member of the Doushantuo Formation, represented by thin-bedded carbonaceous shale with poor solubility, which belongs to asaquiclude. On the way to the Heshangdong Cave from the sinkhole, we can see highly-developed bedding small karst cave, which is closely related to the water-proof fourth member of the Doushantuo Formation.

The rock bearing the sinkhole has developed fractures along SE direction, which became the initial channel of infiltration corrosion for surface water. The sinkhole expanded along the fracture in its formation, and developed into a narrow-top and wide-bottom architecture. Since the dolomite has a poor solubility, the expansion of the cave mainly hinges on the gravitational collapse of its wall, and thus develops collapse deposits in form of the stacking structure. Moreover, the cavity wall, with many fresh rocks of bed rocks exposed, is precipitous and rough.

12.3.2.2 *Observation of the deposits in the cave*

There are meters-thick deposits, with bothcollapse deposits and alluvial deposits brought by the depression stream into the cave. On the north portion, there are many colluvial deposits in form of stacking structure, which are crammed with fine alluvial deposits, forming a gentle slope of meters above the ground on the south portion. As there is no standing water on the bottom, it is postulated that potential groundwater seepage system could be present.

12.3.2.3 *Slide*

Slide refers to the downward displacement of cliff and scarp along the nearly vertical fractured planes. Typically, the vertical displacement is larger than the horizontal movement. The slide structure is well-preserved in the eastern part of the Heshangdong Cave and its trailing edge is characterized by close-to-vertical (about 70°) slide cliff.

Slide differs from other gravitational geological hazards on slope, for instance, landslide contains with typical sliding plane, collapse has high movement speed; whilst slide has tiny displacement with slow speed. The slide would be more likely encountered when the firm rocks are underly by weakly layers, or joints and fractures in the free space have the same directions with the slop.

12.3.3 The project on controlling the potential unstable rocks in Guancaiyan

12.3.3.1 *Formation mechanism of the potential unstable rocks in Guancaiyan*

The top of Gauncaiyan has an elevationat about 827.3 m and the road along the base of the cliff is about 675 m in height, which forms a steep cliff of about 160 m above sea level. There are several factors for the development of potential unstable rocks. Firstly, the cliff has an 80-degree three-side hanging rock body and the south-side slope is even counter-inclined. Secondly, the rocks are intertwined with hard and soft lithologies, with the slope toe consisting of thinly silicic and carbonaceous shales of the fourth member of the Doushantuo Formation that are vulnerable to weathering. Thirdly, due to the coal mining activities, the overlying unstable rocks has developed several slope-paralleled stress-release cracks (Figure 12-2), which form large-scale goaf just below the unstable rocks. This further causes the instability of the rocks. Large falling rocks can also be found along the slope, posing severe threats to the paths and houses.

Figure 12-2 Potential unstable rocks in Guancaiyan developed several slope-paralleled stress-release cracks

12.3.3.2 *Prevention and treatment project of the potential unstable rocks*

Rock-supporting is adopted to measure the potential unstable rock, which includes casting supporting wall with reinforced concrete with steel on the outer edge of the goaf, supporting the roof by casting the concrete column in the inside part, protecting the slopes by concrete gunniting, and installing drainage ditches at the bottom. Those approaches are well-adopted in the prevention and treatment of the small-scale unstable rocks developed along the roads at Huangniuyan. Another useful method is cutting, which means to explode the rock downwards step by step, so as to form a rather stable stepped slope, and thus, it can keep a long-lasting performance. This approach is commonly-seen in the high-speed railway running through the dam, the shipping lock of the dam and the watercourse of the Yangtze River. For large potentially dangerous cliff, like Lianziya Cliff, explosions will probably sprawl potential threats to shipping vessels. Therefore, prevention approaches like anchoring rope and anchoring stock, in spite of higher cost, will be adopted.

12.3.4 Sedimentary characteristics of the Quaternary eluvium talus in Datang

12.3.4.1 *Sedimentary features of the Quaternary eluvium talus*

The studied section locates at the fork of the residential area of Datang, 300 m in the west of Guancaiyan. The section is about 4.5 m high, covered by limited plants, and has been divided into four horizons based on the lithology and color of the sediments. A top–down viewing will make us understand the characteristics of the layers (Figure 12-3).

①yellowish-brown chaotic breccia bed; ②yellowish-brown conglomerate bed;
③grayish-yellow nixed sand and conglomerate bed; ④weathered layer.

Figure 12-3 The outcrop section of the deposits of eluvium in Datang, 300 m south of Guancaiyan

(1) Yellowish-brown chaotic breccia bed. The gravels display unitary lithology of dolomites from the third member of the Doushanduo Formation. Gravels, which are poorly sorted, vary in sizes, with the maximum diameter being about one meter. Gravels are also poor in roundness, most of which are in angular shapes. The A-B faces of gravels are badly oriented. Those gravels are commonly stacked and without cementation, with interspaces fulfilled by fine-grained conglomerates, sands, and muds. This horizon is about one-meter thick and barely weathered.

(2) Yellowish-brown conglomerate bed. Conglomerates in this interval are mainly dolomites which are derived from the third member of the Doushanduo Formation, and

consistent with the bed rocks of the slop. These conglomerates are well sorted, with sizes clustered at 2–15 cm. The A-B faces of the conglomerates slightly incline along the slope; most of the conglomerates display bad roundness, representing by angular shapes. The stacked conglomerates are not cemented, and are fulfilled with fine-grained conglomerates, sands, and muds between the conglomerates. This horizon is about 1.8–2.0-meter thick and barely weathered.

(3) Grayish-yellow mixed sand and conglomerate bed. This horizon is characterized by the chaotic stack of mud-sand and stone blocks. These conglomerates consist of dolomites from the third member of the Doushanduo Formation, and are well-sorted, of sizes at 10–20 cm; but they have poor roundnesses, representing by angular shapes, and are randomly oriented. A 6–8-centimeter thick dark grey muddy sandstone layer can be seen at the top of this horizon, which is undulatedly distributed along the section. Totally, this horizon is about 1.2–1.5 m, but the base is undetectable.

(4) Weathered layer. This horizon is the semi-weathered layer of the second member in the Doushanduo Formation, with a thickness about 1.5–2.0 m.

12.3.4.2 *Cause analysis*

The composition of the deposits and the sedimentological characters of the outcrop suggest that the first layer is formed by slope gravity, the second horizon is drift bed, and the third interval could be the accumulation of debris flow.

12.3.4.3 *Age of the deposits*

These sediments are postulated to be deposited during Late Pleistocene, based on regionally Quaternary stratigraphic correlation.

12.3.5 Impacts of sheet flood erosion and soil-water conservation at Zhangjiachong

12.3.5.1 *Outlook of the small watershed at Zhangjiachong*

The Zhangjiachong area is located at the affluent of the Maopinghe River, which is about 5 km away from the dam and about 8.5 km away from Maoping Town, Zigui. The mean annual precipitation here is about 1,007 mm, with the raindrops predominantly shedding in months between May and October. The rainfall during this period accounts for about 78% of the year total. The average temperature here is about 17.9 ℃. The Zhangjiachong drain basin is typically a closed watershed, with the highest and lowest elevation at 530 m and 148 m respectively.

12.3.5.2 *Conservation of the small watershed in the Zhangjiachong area*

The conservation of the small watershed in the Zhangjiachong area is divided into two parts: the valley and the brae conservation, which can be observed by students.

12.3.5.3 *Observation of soil-water conservation in the experimental station at Zhangjiachong*

The water-soil erosion in the Three Gorges area is mainly due to the sheet wash of the planar flow. The Zhangjaichong watershed area is characterized by wide distributed granites, which are easily to be weathered into quartzose sandy soil, with excellent water permeability and penetrability. In order to find out the pattern of the water-soil erosion in the granite area, the Water and Soil Conservation Bureau of Zigui County has set up Water and Soil Conservation Station in the small watershed of the Zhangjiachong area in September 2002.

The experimental area is set up on the northwest slope of the exit of the watershed, consisting of slope unit and reservoir. The slope unit accounts for about 20 m^2, extending over 11 m and having a slope incline of about 10–25 degrees in a shape of rectangular (Figure 12-4). Under each slope unit, there is an observational pool with a volume of about 3.5 m^3, with which we can record the flow volume of each small area. There are five slopes each with five steps, which are built up with rocks and soil, while the other five slope units are of the same slope gradient. Additional four slopes are set up, whose inclines are 5°, 8°, 15°, and 20° respectively (Figure 12-5).

Figure 12-4 Soil-water conservation and slope layout of the experimental station at Zhangjiachong

Figure 12-5 Experimental tanks for slope soil-water erosion

Some basic designs and distributions include stone-wall ladder and slope hedge for crop, soil-wall ladder ridge hedge for tangerine and slope hedge, and sloping fields for crop, tea, and tangerine, as well as slope wasteland. The major observation points include rainfall, erosion amount, runoff volume, erosion modulus, composition of soil, organic carbon, nitrogen phosphorus content, loss amount, etc.

12.4 Teaching Process and Precautions

12.4.1 Teaching process

(1) Half a day is allocated to the route and five points should be observed. Students are required to do the on-site profile sketch.

(2) Given the fact that we have already learned the stratigraphic route, no more information of bedrock lithology will be explained. Students are required to preview the categories of the Quaternary deposits and the approaches to analyze their causes.

(3) If time is available, on the way to Zhangjiachong, students can observe the alluvial deposits formed in the fluvial process and learn about the methods in measuring the sizes of conglomerates.

(4) During the teaching process, it should be emphasized that the Quaternary sediments are typically unconsolidated, which fundamentally differ from the older strata and are the key points to observe and describe.

12.4.2 Precautions

(1) Considering the winding skyline drive, every teacher and student should be cautious about safety and mind the potential falling rocks when walking under a rock.

(2) Plan in advance how to avoid the danger when it occurs.

12.5 Focused Study and Reflections

(1) Can human activities be seen as an agent? If so, can it be comparable to the outer dynamic geological process?

(2) Analyze the advantages and disadvantages of the Three Gorges Dam Project in an all-round manner and figure out the long-standing and short-term impacts respectively. How can we cope with those impacts?

(3) How is the status quo of the water-soil erosion in the Three Gorges Reservoir area? What are its impacts on the sediment accumulation, eutrophication, soil degradation, and agricultural reclamation? How should we deal with those impacts?

III

ASSESSMENT

13 ASSESSMENT OF TEACHING PROCEDURES AND PRACTICE PERFORMANCES

In line with the objectives of Zigui field teaching and following the requirements of its curriculum outline, this session goes into details about the division of field practice stages, the main teaching contents of each stage, the summaries and key points of each stage, the examination of each stage, the teaching outcome presentation, and the teaching quality inspection and standards of final performance assessment. All these descriptions are made to help teachers and students to fully understand the nature, objectives, and tasks of the field teaching in Zigui, so as to ensure its orderly progress.

13.1 Objectives and Stage Division

13.1.1 Objectives of the internship

Zigui field teaching is conducted on the basis of preceding field trip of comprehensive geological survey in Zhoukoudian. Based on the principle of combining theory and practice, students will systematically figure out the strata similarities and differences between the Yangzi Plate and the North China Plate, which develops their abilities in geological science and make them proficient in geological skills. This paves the way for students' future geological research work.

Zigui field teaching focuses on cultivating students' ability to combine theory with practice to solve specific scientific problems. The training of students' geological thinking ability is always put in the first place, and efforts are made to make students have the consciousness of pioneering, innovative, and scientific research, which is conducive to the teaching of follow-up courses and the development of new talents. At the same time, Zigui field teaching focuses on improving the overall quality of students, cultivating students' spirit of being rigorous and realistic, striving for progress, unity, and cooperation, following organizational discipline, and rising to hardships, which fully reflects the tradition and characteristics of CUG.

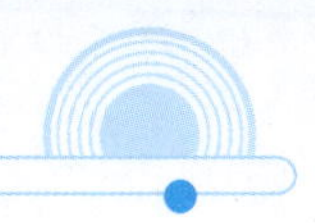

13.1.2 Division of field studies

Zigui field teaching practice lasts for two weeks and can be divided into the following four stages.

(1) Mobilization and preparation for field trip studies. (1 day)

(2) Cognitive teaching (geological teaching on ten routes). (7 days)

(3) Independent research on focused topics. (3 days)

(4) Report preparation on focused topics and assessment. (3 days)

13.2 Main Teaching Contents and Teaching Requirements

13.2.1 Internship mobilization and preparation

The specific contents of the preparation are carried forward by the group leaders and the instructors, including:

(1) The instructor should guide the students to study the field practice outline carefully, and clarify the objectives, tasks, and requirements of each stage, as well as the main teaching contents, teaching points, and scoring standards of assessment. The instructor should highlight practice rules and regulations and precautions, in particular, the demands for study, safety, and confidentiality.

(2) The group leader introduces in detail the geographical situation, the regional geological background, and the present situation of geological research of the field, so that students can have a basic understanding of the area. The instructor should understand the students' degree of mastery on the basic on-field geological skills which have acquired in Zhoukoudian, so as to make the corresponding arrangements.

Instructors carefully check the students' preparation for equipment, instruments, materials, and personal stationery supplies. These include:

(1) Equipment for field studies, such as work clothes, hiking shoes, straw hats, and water bottles.

(2) Personal field practice supplies, such as geological backpack, hammer, magnifiers, compass, field trip log, practice guidance, pencil, knife, and rubber.

(3) Group and personal first aid medicines, etc. Instructors should closely check students' equipment. Those who do not have prepared all necessary equipment cannot take part in field practice.

13.2.2 Cognitive teaching (Teaching of geology on ten routes)

This stage holds the key to the entire field teaching practice. Under the guidance of the instructors, students should do the following tasks by observing geological phenomena on the ten routes.

(1) Students are required to further improve their proficiency in field practice skills and strengthen training in such fields as identification, description, and naming of mineral-and-rock hand samples, as well as identification and division of lithostratigraphic units in the field. They should also pay attention to the standardization of recording format, and sketching of the profiles and typical geological phenomenon.

(2) Students should understand the characteristics and geological significance of ultramafic–mafic igneous assemblage in the field; the distribution, identification characteristics, period of intrusion of intermediate–acid igneous rocks, and its significance on the engineering of the Three Gorges Dam.

(3) Students should master Paleoproterozoic–Mesozoic stratigraphic sequences in the fieldwork area and the age of stratigraphic units, lithologic characteristics, contact relationship, and sedimentary environment of each stratigraphic unit. Thus, they can compare and analyze the correlation of stratigraphic units in North China Block.

(4) Students should understand the petrology, paleontology, and chronology standards established in the GSSP ("Golden Spike") sections.

(5) Students should understand the typical structural types, their characteristics, and evolutions.

In this stage, the teaching schedule is closely arranged, and there is much to be observed in the field, among which some routes being allocated with more time. In this case, students need to carry forward with the spirits of CUG, namely, hard working and persistence in surmounting difficulties. For teachers, they need to make innovations in teaching methods to ensure that teaching tasks are well-completed for follow-up field studies on focused topic.

In this stage's teaching process, the following points should be noticed:

(1) Give full play to students' initiative. In the field-route teaching activities, students are the ones who observe, record, and cognize geological phenomena. Teachers should choose appropriate teaching methods to fully engage students' passion, stimulate students' cognitive enthusiasm and thus, enhance teaching effects. In field teaching, teachers and students are encouraged to discuss more about the geological phenomena, so as to strengthen student-oriented interactions.

(2) Strike a balance between basic teaching content and additional content. Field teaching activities have a certain degree of flexibility. On the basis of ensuring that the basic

teaching content of the syllabus can be understood and accepted by students, teachers can appropriately add some extension and improvement content according to the actual acceptance ability of students and their own research interests. All these efforts are made for students' geological thinking training from point to surface and from micro to macro.

(3) Strengthen the process control of teaching quality. In field teaching, teachers should pay attention to students' cognitive situation, find out their problems in a timely manner when they are observing, recording, describing, and recognizing geological phenomena and correct them on the spot. Teaches should check the field records every day after returning from the field, review the problems, and explain them next day.

13.2.3 Independent focused research

In this stage, students choose the research topic they are interested in, and form a research team of 3–5 people on their own, based on their full understanding of the igneous rocks, stratigraphic sequences, tectonic styles, and other geological features of the internship area. The field data collection work lays the foundation for the writing of research reports on focused topics.

In the stage of independent project research on focused topics, the instructor should respect the students' topic selection and give guidance. After selecting the research content of the topic, students should formulate a more detailed research plan in groups and clarify the division of labor among the members of the group. The instructor should rationally revise their research plans according to the actual abilities of the students in each group to ensure that they can complete relevant fieldwork within a limited time. The research plan of each group needs to be confirmed by the instructor before the work can be carried out. At this stage, teachers should pay special attention to the cultivation of students' scientific research awareness, innovation ability, and teamwork ability.

13.2.4 Report preparation on focused topics and assessment

This stage is the summary of the teaching content and the evaluation of students' study. Reports should focus on cultivating students' ability to process and summarize field data. Students should be able to use basic geological theory and combine first-hand data to make reasonable geological induction and deduction. They should also be able to synthesize various data to organize materials and write reports in an orderly and logical manner. In terms of report format, students can be encouraged to write in the form of scientific research papers. Internship examinations are mainly indoor examinations, focusing on testing students' mastery of field route knowledge points and their independent thinking on research topics. The assessment method is an interview, and the assessment content is flexibly mastered by the instructors in the class, such as discussion, questioning or physical

identification. Teachers will give comprehensive scores according to students' mastery of basic knowledge and independent thinking ability. Those who fail to meet the standards should take effective measures to make up lessons in time.

13.3 Performance Assessment

The evaluation of field practice results is mainly composed of the results of the report on focused study and the practice examination. In addition, students' learning attitude, mastery of basic knowledge, thinking ability, and the quality of field-book records during the field teaching can also be considered.

Instructor registers grades according to the designated form. After weighing into all factors, instructor grades students into five levels, namely outstanding (90–100), good (80–89.5), medium (70–79.5), pass (60–69.5), and fail (0–59.5). In principle, outstanding students in each class shall not exceed 15% of the total number. After synthesizing the scores of each class, the practice team will adjust the balance, evaluate the teaching practice, and report the scores to the Academic Affairs Office and the departments where the students are from.

REFERENCES

白瑾，戴凤岩，1994. 中国早前寒武纪的地壳演化[J]. 地球学报(3/4)：73-87.

陈辉明，孟繁松，张振来，2002. 鄂西秭归盆地下侏罗统桐竹园组新型剖面的研究[J]. 地层学杂志，26(3)：187-192.

陈旭，戎嘉余，樊隽轩，等，2006a. 奥陶系上统赫南特阶全球层型剖面和点位的建立[J]. 地层学杂志，30(4)：289-305.

陈旭，戎嘉余，樊隽轩，等，2006b. 奥陶系—志留系界线地层生物带的全球对比[J]. 古生物学报，39(1)：100-114.

陈旭，袁训来，2013. 地层学与古生物学研究(华南野外实习指南)[M]. 合肥：中国科学技术大学出版社.

DAVIS G A，郑亚东，2002. 变质核杂岩的定义、类型及构造背景[J]. 地质通报，21(4/5)：185-192.

富公勤，袁海华，李世麟，1993. 黄陵断隆北部太古界花岗岩-绿岩地体的发现[J]. 矿物岩石，13(1)：5-13.

范嘉松，1996. 中国生物礁与油气[M]. 北京：海洋出版社.

高山，张本仁，1990. 扬子地台北部太古宙 TTG 片麻岩的发现及其意义[J]. 地球科学(中国地质大学学报)，15 (6)：675-679.

郭俊锋，李勇，舒德干，等，2010. 湖北宜昌纽芬兰统岩家河组结核的特征及形成过程[J]. 沉积学报，28(4)：676-681.

湖北省区域地质测量队，1984. 湖北省古生物图册[M]. 武汉：湖北科学技术出版社.

花友仁，1995. 扬子板块的地壳演化与地层对比[J]. 地质与勘探，31(2)：15-22.

姜继圣，1986. 鄂西黄陵变质地区崆岭群时代及特征的新认识[J]. 长春地质学院学报(1)：100.

赖旭龙，孙亚东，江海水，2009. 峨眉山大火成岩省火山活动与中、晚二叠世之交生物大灭绝[J]. 中国科学基金(6)：353-356.

李长安，殷鸿福，陈德兴，等，1999. 长江中游的防洪问题和对策——1998 年长江特大洪灾的启示[J]. 地球科学(中国地质大学学报)，24(4)：329-334.

李福喜，聂学武，1987. 黄陵断隆北部崆岭群地质时代及地层划分[J]. 湖北地质(1)：28-41.

李清，王家生，陈祈，等，2006. 三峡"盖帽"白云岩中重晶石研究及其古地理意义[J]. 西北大学学报(自然科学版)，36(专辑)：196-200.

李益龙，周汉文，李献华，罗清华，等，2007. 黄陵花岗岩基英云闪长岩的黑云母和角闪石^{40}Ar-^{39}Ar年龄及其冷却曲线[J]. 中国科技期刊研究，23(5)：1067-1074.

李志宏，陈孝红，王传尚，等，2010. 湖北宜昌黄花场下奥陶统弗洛阶上部牙形刺生物地层分带及对比[J]. 中国地质，37(6)：1647-1658.

凌文黎，高山，郑海飞，等，1998. 扬子克拉通黄陵地区崆岭杂岩 Sm-Nd 同位素地质年代学研究[J]. 科学通报，43(1)：86-89.

刘海军，许长海，周祖翼，等，2009. 黄陵隆起形成(165—100Ma)的碎屑岩磷灰石裂变径迹热年代学约束[J]. 自然科学进展，19(12)：1326-1332.

穆恩之，朱兆玲，陈均远，等，1979. 西南地区的奥陶系[M]//中国科学院南京地质古生物研究所. 西南地区碳酸盐生物地层. 北京：科学出版社：108-154.

彭善池，2009. 华南新的寒武纪生物地层序列和年代地层系统[J]. 科学通报，54(18)：2691-2698.

彭善池，2013. 艰难的历程　卓越的贡献——回顾中国的全球年代地层研究[M]//中国科学院南京地质古生物研究所. 中国"金钉子"：全球标准层型剖面和点位研究. 杭州：浙江大学出版社：1-42.

戎嘉余，1984. 上扬子区晚奥陶世海退的生态地层证据与冰川活动影响[J]. 地层学杂志，8(1)：19-29.

戎嘉余，马科斯·约翰逊，杨学长，1984. 上扬子区早志留世(兰多维列世)的海平面变化[J]. 古生物学报，23(6)：672-693，790-792.

沈传波，梅廉夫，刘昭茜，等，2009. 黄陵隆起中—新生代隆升作用的裂变径迹证据[J]. 矿物岩石，29(2)：54-60.

汪啸风，STOUGE S，陈孝红，等，2013. 奥陶系中奥陶统大坪阶全球标准层型剖面和点位及研究进展[M]//中国科学院南京地质古生物研究所. 中国"金钉子"：全球标准层型剖面和点位研究. 杭州：浙江大学出版社：122-150.

汪洋，李勇，张志飞，2011. 峡东水井沱组顶部微体骨骼化石初探[J]. 古生物学报(4)：511-523.

王家生，甘华阳，魏清，等，2005. 三峡"盖帽"白云岩的碳、硫稳定同位素研究及其成因探讨[J]. 现代地质，19(1)：14-20.

王家生，王舟，胡军，等，2012. 华南新元古代"盖帽"碳酸盐岩中甲烷渗漏事件的综合识别特征[J]. 地球科学，37(S2)：14-22.

王永标，2005. 巴颜喀拉及邻区中二叠世古海山的结构与演化[J]. 中国科学(D 辑)，35(12)：1140-1149.

王幼惠，郭成贤，翟永红，1991. 宜昌地区下寒武统沉积环境分析[J]. 江汉石油学院学报，13(3)：15-22.

魏君奇，景明明，2013. 崆岭杂岩中角闪岩类的年代学和地球化学[J]. 地质科学，48(4)：970-983.

魏君奇，王建雄，2012. 崆岭杂岩中斜长角闪岩包体的锆石年龄和 Hf 同位素组成[J]. 高校地质学报，18(4)：589-600.

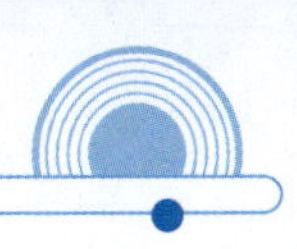

魏君奇，王建雄，王晓地，等，2009. 黄陵地区崆岭群中基性岩脉的定年及意义[J]. 西北大学学报(自然科学版)，39(3)：466-471.

熊成云，韦昌山，金光富，等，1998. 鄂西黄陵背斜核部中段金矿基本特征及成矿规律[J]. 华南地质与矿产(1)：32-40.

熊成云，韦昌山，金光富，等，2004. 鄂西黄陵背斜地区前南华纪古构造格架及主要地质事件[J]. 地质力学学报，10(2)：97-112.

熊庆，郑建平，余淳梅，等，2008. 宜昌圈椅埫 A 型花岗岩锆石 U-Pb 年龄和 Hf 同位素与扬子大陆古元古代克拉通化作用[J]. 科学通报，53(22)：2782-2792.

徐桂荣，罗新民，王永标，等，1997. 长江中游晚二叠世生物礁的生成模型[M]. 武汉：中国地质大学出版社.

尹崇玉，岳昭，高林志，等，1992. 湖北秭归庙河早寒武世水井沱组燧石层中的微化石[J]. 地质学，66(4)：371-380.

袁学诚，1995. 论中国大陆基底构造[J]. 地球物理学报，38(4)：448-459.

喻建新，冯庆来，王永标，等，2016. 三峡地区地质学实习指导手册[M]. 武汉：中国地质大学出版社.

曾庆銮，赖才根，徐光洪，等，1987. 奥陶系[M]//汪啸风，倪世钊，曾庆銮，等. 长江三峡地区生物地层学(2):早古生代部分. 北京：地质出版社：43-142.

张秀莲，于德龙，王贤，2003. 湖北宜昌地区寒武系碳酸盐岩岩石学特征及沉积环境[J]. 古地理学，5(2)：152-161.

郑月蓉，李勇，2010. 三峡地区极短周期内剥蚀速率、下切速率及地表隆升速率对比研究[J]. 成都理工大学学报(自然科学版)，37(5)：513-517.

中国科学院南京地质古生物研究所，2013. 中国“金钉子”——全球标准层型剖面和点位研究[M]. 杭州：浙江大学出版社.

周琦，杜远生，王家生，等，2007. 黔东北地区南华系大塘坡组冷泉碳酸盐岩及其意义[J]. 地球科学，32(3)：339-346.

朱茂炎，2010. 动物的起源和寒武纪大爆发:来自中国的化石证据[J]. 古生物学报，49(3)：269-287.

BAO H，LYONS J R，ZHOU C，2008. Triple oxygen isotope evidence for elevated CO_2 levels after a Neoproterozoic glaciation[J]. Nature，453：504-506.

BASSETT D A，WHITTINGTON H B，WILLIAMS A，1966. The stratigraphy of the Bala district，Merionethshire[J]. The Quarterhy Journal of Geological Society of London，122：219-271.

CAO W C，FENG Q L，FENG F B，et al.，2014. Radiolarian *Kalimnasphaera* from the Cambrian Shuijingtuo Formation in South China[J]. Marine Micropaleontology，110：3-7.

DENG H，KUSKY T M，WANG L，et al.，2012. Discovery of a sheeted dike complex in the northern Yangtze craton and its implications for craton evolution. Journal of Earth Science，23(5)：676-695.

FIELDING C R, 2006. Upper flow regime sheets, lenses and scour fills: extending the range of architectural element for fluvial sediment bodies[J]. Sedimentary Geology, 190: 227-240.

FLÜGEL E, REINHARDT J W, 1989. Uppermost Permian reefs in Skyros (Greece) and Sichuan (China): implications for the Late Permian extinction event[J]. Palaios, 4(6): 502-518.

FAN G H, WANG Y B, KERSHAW S, et al., 2014. Recurrent breakdown of Late Permian reef communities in response to episodic volcanic activities: evidence from southern Guizhou in South China[J]. Facies, 60: 603-613.

HOFFMAN PF, KAUFMAN A J, HALVERSON G P, et al., 1998. A Neoproterozoic snowball Earth[J]. Science, 281(5381): 1342-1346.

HU J, WANG J S, CHEN H R, et al., 2012. Multiple cycles of glacier advance and retreat during the Nantuo (Marinoan) glacial termination in the Three Gorges area[J]. Frontiers of Earth Science, 6(1): 101-108.

INGHAM J K, WRIGHT A D, 1970. A revised classification of the Ashgill Series[J]. Lethaia, 3: 233-242.

JIANG G, KENNEDY M J, CHRISTIE-BLICK N, 2003. Stable isotopic evidence for methane seeps in Neoproterozoic postglacial cap carbonates[J]. Nature, 426: 822-826.

WANGJ S, JIANG G Q, XIAO S H, et al., 2008. Carbon isotope evidence for widespread methane seeps in the ca. 635 Ma Doushantuo cap carbonate in south China[J]. Geology, 36(5): 347-350.

KENNEDY M J, CHRISTIE-BLICK N, SOHL L E, 2001. Are Pretorozoic cap carbonates and isotopic excursions a record of gas hydrate destablilization following Earth's coldest intervals?[J]. Geology, 29(5): 443-446.

LIU S F, ZHANG G W, 2013. Mesozoic basin development and its indication of collisional orogeny in the Dabie orogen[J]. Chinese Science Bulletin, 58(8): 827-852.

LIU S F, QIAN T, LI W P, et al., 2015. Oblique closure of the northeastern Paleo-Tethys in central China[J]. Tectonics, 34(3): 413-434.

LIU S F, STEEL R, ZHANG G, 2005. Mesozoic sedimentary basin development and tectonic implication, northern Yangtze Block, eastern China: record of continent-continent collision[J]. Journal of Asian Earth Sciences, 25(1): 9-27.

MAILL A D, 1985. Architectural-element analysis: a new method of facies analysis applied to fluvial deposits[J]. Earth Science Review, 22: 261-308.

MCFADDEN K A, HUANG J, CHU X L, et al., 2008. Pulsed oxidation and biological evolution in the Ediacaran Doushantuo Foramtion[J]. Proceedings of the National Academy of Sciences of the United States of America, 105(9): 3197-3202.

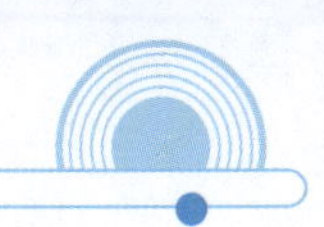

SHE Z B, MA C Q, WAN Y S, et al., 2012. An Early Mesozoic transcontinental palaeoriver in South China: evidence from detrital zircon U-Pb geochronology and Hf isotopes[J]. Journal of the Geological Society, 169(3): 353-362.

SHEEHAN P M, 2001. The Late Ordovician mass extinction[J]. Annual Reviews of Earth and Planetary Sciences, 29(1): 331-364.

WANG Y D, 2002. Fern ecological implications from the Lower Jurassic in Western Hubei, China[J]. Review of Palaeobotany and Palynology, 119(1/2): 125-141.

WANG X F, STOUGE S, ERDTMANN B D, et al., 2005. A proposed GSSP for the base of the Middle Ordovician Series: the Huanghuachang section, Yichang, China[J]. Episodes, 28(2): 105-117.

WIGNALL P B, SUN Y D, BOND D P G, et al., 2009. Volcanism, mass extinction, and carbon isotope fluctuations in the Middle Permian of China[J]. Science, 324: 1179-1182.

WIGNALL P B, 2001. Large igneous provinces and mass extinctions[J]. Earth-Science Reviews, 53: 1-33.

XIAO S H, ZHANG Y, KNOLL A H, 1998. Three-dimensional preservation of algae and animal embryos in a Neoproterozoic phosphorite[J]. Nature, 391: 553-558.

YANG J H, CAWOOD P A, DU Y S, 2010. Detrital record of mountain building: provennace of Jurassic foreland basin to the Dabie Mountains[J/OL]. Tectonics, (2010-07-27)[2014-05-20]. https://agupubs.onlinelibrary.wiley.com/doi/full/10.1029/2009TC002600.

ZHEN Y Y, LIU J B, PERCIVAL I G, 2005. Revision of two Prioniodontid species (Conodonta) from the Early Ordovician Honghuayuan Formation of Guizhou, South China [J]. Records of the Australian Museum, 57: 303-320.